Drones for Biodiversity Conservation and Ecological Monitoring

Drones for Biodiversity Conservation and Ecological Monitoring

Special Issue Editors

Ricardo Díaz-Delgado
Sander Mücher

MDPI • Basel • Beijing • Wuhan • Barcelona • Belgrade

MDPI

Special Issue Editors

Ricardo Díaz-Delgado
LAST (Remote Sensing & GIS Lab),
Do ñana Biological Station-CSIC
Spain

Sander Mücher
Wageningen Environmental Research
(Alterra) Wageningen Campus
The Netherlands

Editorial Office
MDPI
St. Alban-Anlage 66
4052 Basel, Switzerland

This is a reprint of articles from the Special Issue published online in the open access journal *Drones* (ISSN 2504-446X) from 2018 to 2019 (available at: https://www.mdpi.com/journal/drones/special_issues/biodivers).

For citation purposes, cite each article independently as indicated on the article page online and as indicated below:

LastName, A.A.; LastName, B.B.; LastName, C.C. Article Title. *Journal Name* **Year**, *Article Number*, Page Range.

ISBN 978-3-03921-980-3 (Pbk)
ISBN 978-3-03921-981-0 (PDF)

Contents

About the Special Issue Editors

Ricardo Díaz-Delgado (PhD) received his BS degree in Biology in 1994 from UAM, a MS degree in Mediterranean Ecosystems in 1996 from the IMEP-Université d'Aix-Marseille III, a MS degree in Remote Sensing Applications from the Institute for Space Studies of Catalonia (Barcelona) in 1997, and a PhD degree in Application of Remote Sensing to Fire Ecology from the Autonomous University of Barcelona in 2000. Since 2001, he has worked at the Doñana Biological Station sited in Sevilla, leading the Remote Sensing and GIS Laboratory. His research focuses on the application of remote sensing and spatial information systems to ecological monitoring, especially of disturbance processes and biological conservation.

Sander Mücher (PhD) is senior researcher Remote Sensing and Geo-Information at Wageningen Environmental Research (WENR), within the team Earth Observation and Environmental Informatics. He has been active for more than 25 years at WENR in the field of remote sensing and the application domains of biodiversity, agriculture, and environmental monitoring. In 1997 he was already co-ordinator of the EU-FP4 project PELCOM which aimed at European land use monitoring with satellites for environmental applications. Together with Wageningen University, he established, in 2012, a ROC-certified unmanned aerial remote-sensing facility (WUR-UARSF) with a wide range of national and international applications in the field of agriculture and nature (http://www.wur.eu/uarsf). He is now involved in many European remote sensing activities, particularly vegetation mapping and monitoring at different scales for agriculture and biodiversity. He has been involved in many European projects such as BIOPRESS, BIOHAB, EBONE, ECOCHANGE, SIGMA, and BIO-SOS. Within the current Interrreg project SPECTORS, he is responsible for developing UAS services for nature monitoring, and in the H2020 project GENTORE he is aiming at the exploitation of drones and artificial intelligence to aid in the identification and characterization of animals. Next to these international activities, he is also involved in many national projects in the field of hyperspectral and LiDAR-based UAS applications for biodiversity and phenotyping. The integration of remote sensing and in-situ information for monitoring and modelling plays an important role in all his studies.

drones

MDPI

Editorial

Editorial of Special Issue "Drones for Biodiversity Conservation and Ecological Monitoring"

Ricardo Díaz-Delgado [1,*] **and Sander Mücher** [2]

[1] Remote Sensing and GIS Laboratory (LAST-EBD). Estación Biologica de Doñana. CSIC. Avda. Américo Vespucio 26, 41092 Sevilla, Spain
[2] Wageningen Environmental Research (WENR), Wageningen University and Research, Building 101, Droevendaalsesteeg 3, P.O Box 47, 6700 AA Wageningen, The Netherlands; sander.mucher@wur.nl
* Correspondence: rdiaz@ebd.csic.es; Tel.: +34-954-232340

Received: 2 June 2019; Accepted: 3 June 2019; Published: 7 June 2019

Abstract: Unmanned Aerial Vehicles (UAV) have already become an affordable and cost-efficient tool to quickly map a targeted area for many emerging applications in the arena of Ecological Monitoring and Biodiversity Conservation. Managers, owners, companies and scientists are using professional drones equipped with high-resolution visible, multispectral or thermal cameras to assess the state of ecosystems, the effect of disturbances, or the dynamics and changes of biological communities inter alia. It is now a defining time to assess the use of drones for these types of applications over natural areas and protected areas. UAV missions are increasing but most of them are just testing its applicability. It is time now to move to frequent revisiting missions, aiding in the retrieval of important biophysical parameters in ecosystems or mapping species distributions. This Special Issue is aimed at collecting UAV applications contributing to a better understanding of biodiversity and ecosystem status, threats, changes and trends. Submissions were welcomed from purely scientific missions to operational management missions, evidencing the enhancement of knowledge in: Essential biodiversity variables and ecosystem services mapping; ecological integrity parameters mapping; long-term ecological monitoring based on UAVs; mapping of alien species spread and distribution; upscaling ecological variables from drone to satellite images: methods and approaches; rapid risk and disturbance assessment using drones, ecosystem structure and processes assessment by using UAVs, mapping threats, vulnerability and conservation issues of biological communities and species; mapping of phenological and temporal trends and habitat mapping; monitoring and reporting of conservation status.

Keywords: UAVs; ecological monitoring; biological conservation; drone mapping; biodiversity; phenology

---

1. Special Issue Overview

This special issue was proposed in 2017, November the 14th, and kindly accepted on November the 21st. Announcements were spread during January 2018 through many different mailing lists dealing with remote sensing and Geographic Information System, as well as personal invitations to a list of 232 authors with previous publications, on the use of UAVs in relation to the topics. First abstracts were then submitted immediately after the call, having the first accepted and published paper on 12th April. The initial deadline was scheduled for 31st July, but it was decided to extend to 31st October.

A total amount of 12 manuscripts were submitted from which 10 were finally accepted and published, including one review paper and 11 research articles. All of them were peer-reviewed by scientific reviewers suggested by authors and the editorial board. Some of the manuscripts had to prepare full re-submissions according to reviewer's suggestions. Rejections were merely related to be out of scope. MDPI gently allowed submission of manuscripts which had been previously compromised by authors with abstracts until November 2018. While the Special Issue was open, flyers were kindly prepared by MDPI which were distributed in several conferences and research project meetings.

2. Special Issue Topics

A literature search using ISI Web of Science with the option, TS (UAV or UAS or Drone), provided 28,485 records from which 30% were classified under, engineering electrical electronic, 13% under automation control systems and in the 3d position, remote sensing was found. The publication year starts in 2010 with 720 publications and increases to 4723 in 2018. The USA leads the academic production followed by China. By adding, and TS (REMOTE SENSING), records drop down to a total of 1860, with just 37 indexed papers in 2010 to 475 in 2018. From these, 782 are strictly classified under the remote sensing category and just 161 under the Environmental sciences category. This quick literature search points out the delay in the application of drones for environmental issues in relation to other knowledge fields. Additionally, it reveals the lack of published papers in this area.

This special issue has contributed to the spread the many diverse applications of drones in ecological monitoring and biodiversity conservation. Three of the published articles have addressed the main issues and opportunities in wildlife detection and mapping, and in both cases applied to protected species. Bonnin et al. [1] used fixed-wing drones by means of in line transects and grid missions to locate Chimpanzee nests in Tanzania. Potential unknown nests were identified although many nests located in the middle of tree crowns could not be detected. The authors suggest the combination of lower altitude flights with multirotor drones, multispectral cameras and oblique surveys can lead to reduce false alerts and increase nest detectability. In any case, this study indicates that one of the most widely used application of drones in wildland management, although most of them remain unpublished. Similarly, Afán et al. [2] address the suitability of drone-borne images to automatically map and count Glossy Ibis nests. While monitoring bird populations and productivity, the staff have to recurrently visit the colony, and once per year enter the colony to confirm the census. This effort is clearly benefited from the accurate counting by drone-borne images which clearly appear as one of the most efficient application in environmental monitoring. Machine learning techniques are contributing to enhance counting by identifying different species. One interesting option to locate wildlife is, rather than feature identification or mapping, is to use drones for radio-tracking. This is the subject addressed by Desrochers et al. [3] in its published paper, where a VHF (Very High Frequency) receiver was mounted in the drone to test the accuracy in locating radio-transmitters in a dense boreal forest. The results demonstrated that drone radio-tracking offers an efficient alternative to the more labor-intensive, traditional approaches for radio-tracking small birds, amphibians, or small mammals in rugged terrain. Further, more essays have to be completed to evaluate the limitations while tracking moving animals and defining adequate flight paths or search patterns to find close signals. All three studies are good examples of the wide spectrum brightly depicted by Jiménez López and Mulero Pázmány [4] in the special issue review. The authors summarize the search for published contributions of drone applications in wildlife monitoring and management, ecosystem monitoring, law enforcement, ecotourism and environmental management and disaster response. An immense diversity of applications reveals the wide flexibility and ingenuity of drone missions to retrieve critical spatial data dealing with environmental monitoring and assessment.

A set of published articles under this special issue point out the important role of drone-borne multispectral images for validation and upscaling of satellite images and derived products. This is the case for the papers by Díaz-Delgado et al. [5], Pla et al. [6] and Canisius et al. [7]. The first paper takes advantage of the high spatial resolution of multispectral images acquired over Doñana marshes to retrieve relevant monitoring indicators related to ecological integrity of the ecosystem. The intermediate scale provided by multispectral drone images are used to enlarge ground-truth for Sentinel-2 images, which exposes a new methodological approach to assess the accuracy of large scale ecosystem essential variables provided by Earth observation satellites, such as the Copernicus European program or the international GEO-GEOSS. Similarly, Pla et al [6], used NDVI (Normalized

Difference Vegetation Index) from multispectral images from a multirotor drone to evaluate damages by one threatened bird species on rice paddies. Fine grain damages accounted from drone images was later related to Sentinel-2 NDVI images in order to extend the assessment to larger areas. Although the study was carried out on agricultural area, such agro-systems are habitats for the wildlife and therefore, payments for damages have to be properly estimated to preserve biodiversity. Finally, Canisius et al. [7], developed a multi-sensor platform to reliably measure land surface albedo providing a strong relationship between direct total shortwave albedo measurements from the pyranometer mounted on the UAV (Unmanned Aerial System) with Sentinel-2 and Landsat-8 estimates. This paper demonstrates how drones can bridge the gap between fixed point, in-situ albedo measurements and pixel-level measurements by satellites.

The special issue has also revisited classical remote sensing approaches to assist in post-fire recovery monitoring and vegetation mapping using drones. Melville et al. [8] presented an excellent example of accurate mapping of grassland communities with the use of a hyperspectral sensor. This technology has quickly become affordable and light enough to be mounted in powerful multirotor drones allowing to retrieve a continuum spectrum of plant spectral signatures. The authors also made use of one of the most advantageous outcomes of drone mapping—point cloud and the derived digital surface models. From this information, always inherently produced by drone gridded missions, canopy height models as a function of bare ground models can also be retrieved. Their classification procedure using random forest was enhanced with this valuable information. Fine scale and accurate habitat mapping and monitoring with drones is evidenced by what becomes a reliable technique as inputs in habitat assessment, such as the reporting on conservation status requested by the Natura 2000 European directive. Larrinaga and Brotons [9] also harnessed the Structure from Motion (SfR) to derive canopy height and together, with several multispectral vegetation indices, use it to model tree DBH (Diameter at the Brest Height) in a 20-year old burned area. The authors predict forest recovery after fire with DBH as an indicator of plant canopy. The last, but not least, contribution by Díaz-Delgado et al. [10] shows a very innovative proposal to use drones in long-term monitoring of experimental climate change plots. The authors stressed the need to complement in situ measurements in order to complement them by adding spatially explicit information to the effect of treatments on native grassland communities. One of the findings reveals that heterogeneity in plant response to treatments may easily hide species specific trends and changes.

There were no contributions of applications using LiDAR sensors onboard of drones which is known to be widely used on natural areas with many different objectives in the framework of ecological monitoring. Researchers and technicians are invited to submit their studies to Drones Open Access journal. The authors look forward to have our journal indexed in the close future.

3. Message to Conservationists, Practitioners, Managers and Ecologists

Nowadays, the joint development of sensors, drones and image analysis techniques as machine learning is providing an excellent opportunity to aid in environmental conservation and management. Furthermore, not only rapid missions to assess the effects of sudden disturbances, but periodical surveys carried out by programmed flights by drones, are providing invaluable assistance in decision making as spatially explicit ecological and biodiversity monitoring. It is a matter of researchers to provide and diffuse the suitability of drones to carry out very specific and timely missions enhancing the necessary labor of day-to-day nature monitoring and management. There is also much discussion about the aviation rules that should enable E-VLOS and B-VLOS to increase the operational use for agriculture and nature applications. On the other hand, more research is needed about disturbances of UAVs on species. The outcome of these discussions can have a big impact on the further use of UAVs. Nevertheless, there is a wide consensus on the upcoming strong increase in the use of drones for these topics.

Acknowledgments: The first assistant editor, Ms. Trista Liu, provided great support to guest editors, Ricardo Díaz-Delgado and Sander Mücher, until her handover to the new editorial officer, Elvis Wang, who dealt with most of the edition of this Special Issue.

References

1. Bonnin, N.; Van Andel, A.C.; Kerby, J.T.; Piel, A.K.; Pintea, L.; Wich, S.A. Assessment of Chimpanzee Nest Detectability in Drone-Acquired Images. *Drones* **2018**, *2*, 17. [CrossRef]
2. Afán, I.; Máñez, M.; Díaz-Delgado, R. Drone Monitoring of Breeding Waterbird Populations: The Case of the Glossy Ibis. *Drones* **2018**, *2*, 42. [CrossRef]
3. Desrochers, A.; Tremblay, J.A.; Aubry, Y.; Chabot, D.; Pace, P.; Bird, D.M. Estimating Wildlife Tag Location Errors from a VHF Receiver Mounted on a Drone. *Drones* **2018**, *2*, 44. [CrossRef]
4. Jiménez López, J.; Mulero-Pázmány, M. Drones for Conservation in Protected Areas: Present and Future. *Drones* **2019**, *3*, 10. [CrossRef]
5. Díaz-Delgado, R.; Cazacu, C.; Adamescu, M. Rapid Assessment of Ecological Integrity for LTER Wetland Sites by Using UAV Multispectral Mapping. *Drones* **2019**, *3*, 3. [CrossRef]
6. Pla, M.; Bota, G.; Duane, A.; Balagué, J.; Curcó, A.; Gutiérrez, R.; Brotons, L. Calibrating Sentinel-2 Imagery with Multispectral UAV Derived Information to Quantify Damages in Mediterranean Rice Crops Caused by Western Swamphen (Porphyrio porphyrio). *Drones* **2019**, *3*, 45. [CrossRef]
7. Canisius, F.; Wang, S.; Croft, H.; Leblanc, S.G.; Russell, H.A.J.; Chen, J.; Wang, R. A UAV-Based Sensor System for Measuring Land Surface Albedo: Tested over a Boreal Peatland Ecosystem. *Drones* **2019**, *3*, 27. [CrossRef]
8. Melville, B.; Lucieer, A.; Aryal, J. Classification of Lowland Native Grassland Communities Using Hyperspectral Unmanned Aircraft System (UAS) Imagery in the Tasmanian Midlands. *Drones* **2019**, *3*, 5. [CrossRef]
9. Larrinaga, A.R.; Brotons, L. Greenness Indices from a Low-Cost UAV Imagery as Tools for Monitoring Post-Fire Forest Recovery. *Drones* **2019**, *3*, 6. [CrossRef]
10. Díaz-Delgado, R.; Ónodi, G.; Kröel-Dulay, G.; Kertész, M. Enhancement of Ecological Field Experimental Research by Means of UAV Multispectral Sensing. *Drones* **2019**, *3*, 7. [CrossRef]

![drones logo] *drones*

MDPI

Article

Calibrating Sentinel-2 Imagery with Multispectral UAV Derived Information to Quantify Damages in Mediterranean Rice Crops Caused by Western Swamphen (*Porphyrio porphyrio*)

Magda Pla [1,*], Gerard Bota [1], Andrea Duane [1], Jaume Balagué [1], Antoni Curcó [4], Ricard Gutiérrez [5] and Lluís Brotons [1,2,3]

[1] InForest JRU (CTFC-CREAF), Carretera de Sant Llorenç de Morunys Km 2, 25280 Solsona, Lleida, Spain; gerard.bota@ctfc.cat (G.B.); andrea.duane@ctfc.cat (A.D.); jaume.balague@ctfc.es (J.B.); lluis.brotons@ctfc.cat (L.B.)

[2] Centre for Ecological Research and Forestry Applications (CREAF), 08193 Cerdanyola del Vallès, Spain

[3] Spanish National Research Council (CSIC), 08193 Cerdanyola del Vallès, Spain

[4] Ebro Delta Natural Park, Government of Catalonia, Av. Catalunya, 46, 43580 Deltebre, Spain; acurcom@gencat.cat

[5] Ministry of Territory and Sustainability, Government of Catalonia, Dr. Roux, 80, 08017 Barcelona, Spain; rgutierrez@gencat.cat

* Correspondence: magda.pla@ctfc.cat; Tel.: +34-973-481-752

Received: 22 March 2019; Accepted: 13 May 2019; Published: 21 May 2019

Abstract: Making agricultural production compatible with the conservation of biological diversity is a priority in areas in which human–wildlife conflicts arise. The threatened Western Swamphen (*Porphyrio porphyrio*) feeds on rice, inducing crop damage and leading to decreases in rice production. Due to the Swamphen protection status, economic compensation policies have been put in place to compensate farmers for these damages, thus requiring an accurate, quantitative, and cost-effective evaluation of rice crop losses over large territories. We used information captured from a UAV (Unmanned Aerial Vehicle) equipped with a multispectral Parrot SEQUOIA camera as ground-truth information to calibrate Sentinel-2 imagery to quantify damages in the region of Ebro Delta, western Mediterranean. UAV vegetation index NDVI (Normalized Difference Vegetation Index) allowed estimation of damages in rice crops at 10 cm pixel resolution by discriminating no-green vegetation pixels. Once co-registered with Sentinel grid, we predicted the UAV damage proportion at a 10 m resolution as a function of Sentinel-2 NDVI, and then we extrapolated the fitted model to the whole Sentinel-2 Ebro Delta image. Finally, the damage predicted with Sentinel-2 data was quantified at the agricultural plot level and validated with field information compiled on the ground by Rangers Service. We found that Sentinel2-NDVI data explained up to 57% of damage reported with UAV. The final validation with Rangers Service data pointed out some limitations in our procedure that leads the way to improving future development. Sentinel2 imagery calibrated with UAV information proved to be a viable and cost-efficient alternative to quantify damages in rice crops at large scales.

Keywords: Sentinel; UAV; Parrot SEQUOIA; multispectral; vegetation indices; rice crops; western swamphen

---

1. Introduction

Addressing human–wildlife conflicts is a fundamental challenge for conservation practitioners. In some areas of the planet, the loss of lives, crops, or live-stock because of wildlife has significant consequences for people's livelihoods and their food and agricultural security [1,2]. Reconciling the conservation of endangered species with human activities, as well as with socioeconomic uses and human security, has been the subject of numerous studies and management programs. The most commonly implemented actions are aimed at mitigating wildlife negative effects on agricultural production and human health through mitigation operations or economical compensations for losses caused by wildlife [3–5]. Quantifying and mapping wildlife-caused damages is essential to carry out both kind of actions. Most methodologies are aimed at developing predictive risk maps [6], but in the context of human–wildlife conflicts, it is crucial to develop accurate protocols for the reliable verification of the authority of the causative species and their relation with damage claims [7]. These protocols are fundamental in creating public trust in the legitimacy of compensation programs, and in avoiding fraud and moral hazards. Finding simple and inexpensive methods to quantify damages is a major challenge in decreasing the cost–benefit ratio to achieve conservation objectives. Medium resolution remote sensing imagery as low-cost Unmanned Aerial Vehicles (UAV) has arisen as an essential tool to meet this challenge.

Rice crops (*Oriza sativa*) are one of the agricultural habitats most affected by human–wildlife conflicts [8,9]. Large populations of aquatic birds inhabit these agricultural habitats due to their high productivity levels and similarity to natural wetlands [3]. Rice crops are one of the most important bases of the rural economy in many regions of the world. For instance, in the region of the Ebro Delta in the western Mediterranean, with more than 20,000 ha of surface area, the rice fields represent an essential economic revenue, accounting for a significant proportion of the Spanish rice production. At the same time, the Ebro Delta constitutes a fundamental area for the conservation of biodiversity. Hence, some protection figures have been established to protect the area, namely the Natural Park and Wildlife Reserve, Wetlands of International Importance (Ramsar Convention), Natura 2000 Network Space, and recently, Biosphere Reserve for UNESCO's Man and the Biosphere Program. In this context, striving for agricultural production compatible with the conservation of the biological wealth of the Delta is a priority objective of public administrations and represents an important challenge for landscape managers.

The Western Swamphen is a species of conservation interest in Europe that is included in Annex I of the Birds Directive and is considered a Least Concern species at the European level [10]. The population in Europe is estimated at 3400–3800 pairs [10]. At the moment, the Iberian Peninsula contains 81% of this population [10], and thanks to the application of conservation measures, at the end of twentieth century the species had naturally recolonized some of the Iberian wetlands where it had disappeared [11]. However, its population increase in the Ebro Delta has caused the emergence of conflicts with rice farmers, since the Western Swamphen uses rice crops as feeding grounds, leading to crop losses. Since the Western Swamphen is protected by national and European legislation, actions taken to reduce crop damages that may lead to changes in its conservation status, such as lethal control or captures and translocations, cannot be undertaken without an assessment of alternative measures. In this context, objectively quantifying Swamphen damage as accurately as possible is a requisite. Initially, the damage assessment was only carried out based on field visits by regional ranges to the affected crops. The obvious damages near the roads were easily evaluable and quantifiable, but the less obvious damages located further away from the roads remained difficult to evaluate. In addition, this method may be subjected to certain levels of subjectivity and without accurate evaluation of the committed error. Thus, defining an objective methodology to quantify the damages is an essential challenge, both for the benefit of the species and from an economic point of view.

Remote sensing medium resolution imagery has shown a great capacity to quantify damages in crop lands. The satellites MODIS and Landsat have been widely used [12–14], but Sentinel-2, especially for its finer resolution, constitutes a major asset for this kind of application [15,16]. However, due to the types of damages that affect the majority of the crops, satellite resolutions are still too coarse. In this sense, imagery provided by the high spatial resolution commercial satellites as Deimos-2, GeoEye-2, QuickBird, or WorldView-2 could be an appropriate option [14,17,18]. Nevertheless, in some specific decision making contexts the cost–benefit ratio is too high, especially if multiple images of different dates are required or large surfaces are needed. The use of lightweight UAV usually implies lower economic costs than other remote sensing techniques when surveying relatively small areas (tens of hectares) and can be a very good alternative [19–24], especially due to its finer resolution (better than 20 cm) and versatility. Lightweight UAVs and photogrammetric derived information as Digital Surface Models (DSM) have been used to calculate crop damages by means of crop height estimation [25] or grapevine canopy changes [26]. UAV multispectral information is also becoming very important due to its ability to asses vegetation stress and damages through vegetation indices [27,28]. In fact, this ability can be very useful to detect Western Swamphen damages, usually resulting in small and medium vegetation gaps and plant vigor decreases in rice crops. However, mapping large areas with UAV, such as the Ebro Delta, with more than 20,000 ha, may require considerable economic and technical efforts [29]. In this context, some works use UAV information as field data to calibrate imagery from coarser resolution satellites, such as Landsat or Sentinel-2, with correlative methodologies [30,31].

In addition to the developments in quantitative methodological work-flows, it is equally important to offer the results in a comprehensible and useful format for managers responsible for damage assessment and economic compensation guidance [6]. Developing comprehensible products to boost conservation strategies at high efficiency ratios is a prerequisite.

Thus, this work has two main objectives: the first one is to test the ability to use UAV multispectral imagery as ground truth information for calibrating the physical parameters calculated on medium resolution satellite remote sensing imagery from Sentinel-2; the second is to translate the calibrated remote sensing quantification into understandable and useful products for the landscape managers. We tested these questions in the Ebro Delta region, in NE Spain, with rice crops affected by Western Swamphen activity.

2. Study Area and Species

The Ebro Delta constitutes one of the largest wetland areas of the western Mediterranean, at 320 km^2. Its location per Ramsar site designation is 40°43′ N, 00°44′ E (Figure 1). The delta protrudes around 30 km seaward, with a triangular shape created by the river. Of this, 20% of the area corresponds to natural habitats, 75% is arable land, and the rest are urban areas. Most of the arable land is used for intensive rice agriculture, but noticeable is that this type of agriculture was not implemented until the late nineteenth century. Natural habitats encompass rich ecosystems corresponding to rivers, sea, bays, beaches, dunes, riparian forests, coastal lagoons, and saltmarshes. As a result, the Ebro Delta was declared a natural park in 1983 and a biosphere reserve in 2013. Rice paddies occupy 65% of the Ebro Delta area, and together with tourism, constitute the main economic resource for the approximately 15,000 local residents [3]. An abundant population of Western Swamphen has been recorded, mainly distributed in natural wetlands and their surroundings, including rice fields [32]. At present, the Iberian Peninsula constitutes most of its European population and at the end of the 20th century there was a significant population growth thanks to the implementation of active conservation measures [11]. Western Swamphen cause rice damage through the cutting of stems for feeding and by trampling during displacement (creating corridors) within the rice field (Figure 2). Also, in the case of individual breeding, plaques are generated in the crop (Figure 2, right) due to the construction of nests and the consequent trampling around them. The effects of Western Swamphen in the field are therefore: (1) absence of plants (stems cut) and (2) damaged plants that rebound (with smaller size and lower rice productivity). Overall, the affected field experiences decreased rice productivity. Due to existing regulation [33], the government of Catalonia (regional administration) compensates the affected farmers according to the surface area damaged. In the Ebro Delta, administrators had paid up to €203,450 per year from 2003 to 2016 to compensate for damages on about 21,000 ha of rice fields [3].

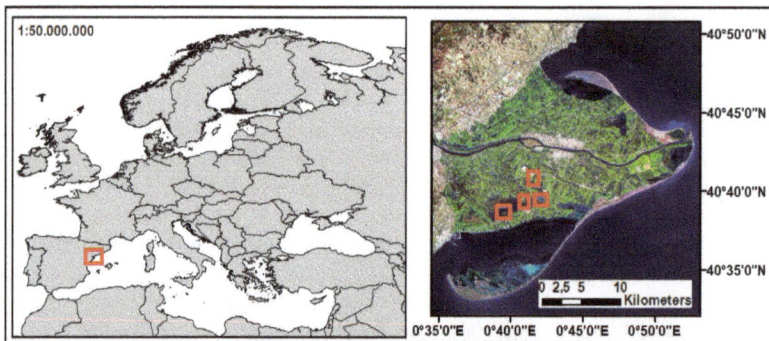

Figure 1. (Left) Ebro Delta location (red rectangle) over a Europe Administrative limits map. **(Right)** Landsat image of the Ebro Delta. Green land cover corresponds to rice crops. Red rectangles show the four agricultural plots where UAV (Unmanned Aerial Vehicle) images were captured (surveyed plots).

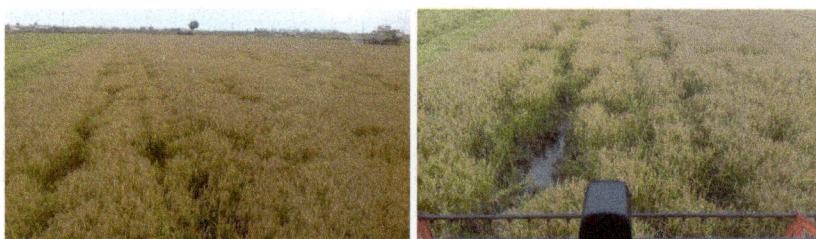

Figure 2. Field photos of damage caused by the western Swamphen on rice plots. **(Left)** Corridors with damaged plants within the rice field, the most abundant damages. **(Right)** A plaque with a lack of vegetation leaving the crop water visible.

3. Material and Methods

We aimed to: (1) calibrate Sentinel-2 images with UAV-derived data, and then (2) validate the generated Sentinel-2 products with field information. The UAV multispectral imagery data were captured in four agricultural plots and used as ground-truth information. The four plots were representative of different levels of crop damage and types of cultivated rice. To assess the matching of information between UAV images and satellite images, we calculated the proportion of damage in each square of a 10 × 10 m mesh co-registered with Sentinel-2 pixels. Then, we selected Sentinel-2 images close to the date of UAV imagery and calculated the Normalized difference vegetation index (NDVI), explained in more detail in the following sections. To reduce noise in the analyses, the 10 m pixels that were not purely of rice crops were removed. We then built a model that related the proportion of damage assessed by UAV imagery to the Sentinel-2 NDVI data. We used the fitted relationship to extrapolate rice damage to the whole Sentinel-2 Delta Ebro image. The prediction maps were validated with independent damage information derived from field observations at the plot level. The following sections explain the entire procedure in more detail.

3.1. Sentinel-2 Imagery

We used free Sentinel-2 imagery that had already been atmospherically-corrected by European Space Agency's (ESA) Sen2Cor algorithm (Level-2A) [34]. We selected sixteen available cloud-free images from April to August 2017 corresponding to the Sentinel-2 31TCF zone: April 3, April 6, May 6, May 16, May 23, May 26, June 12, June 15, June 22, July 2, July 5, July 12, August 4, August 14, August 21, and August 24. We calculated the Normalized Difference Water Index (NDWI) for every date.

NDWI was proposed by McFeeters in 1996 to delineate the open water features by means of the green and NIR wavelengths:

$$NDWI = \frac{GREEN - NIR}{GREEN + NIR} \qquad (1)$$

The water surfaces have positive values, while the surface of the earth and vegetation have 0 or negative values. This NDWI time series was used to identify water channels near the fields to allow later selection of pure rice crops squares, as is explained in Section 3.4. The August 24 image was the one selected to quantify the maximum crop damages, because it matches the highest season of Western Swamphen activity and is the available image close to the capture date of UAV images. Some small clouds and their shadows present in this image were manually removed by screen digitization. We calculated NDVI for the August 24 Sentinel-2 image, and we refer to this as Sentinel2-NDVI.

3.2. UAV Imagery Adquisition

High resolution image data were collected using a quadcopter (UAV Phantom 3 from DJI) [35]. This light weight (1280 g) UAV is capable of autonomous waypoint flight following a preplanned route (Figure 3). A Parrot SEQUOIA multispectral camera was installed in the quadcopter, with four 1.2 megapixel monochrome sensors that collected global shutter imagery along four discrete spectral bands: green (center wavelength (CW): 550 nm; bandwidth (BW): 40 nm), red (CW: 660 nm; BW: 40 nm), red edge (CW: 735 nm; BW: 10 nm), and near infrared (CW: 790 nm; BW: 40 nm). The horizontal (H), vertical (V), and diagonal (D) fields of view (FOV) of the multispectral camera were 61.9° (HFOV), 48.5° (VFOV), and 73.7° (DFOV), respectively, with a focal length of 4 mm. The camera was bundled with an irradiance sensor to record light conditions in the same spectral bands as the multispectral sensor ISO (International Organization for Standardization) value and exposure time was set to automatic. The setting of every captured image is saved in a text metadata file with the irradiance sensor data. All this information was taken into account during the preprocessing stage to obtain reflectance values for the final multispectral product.

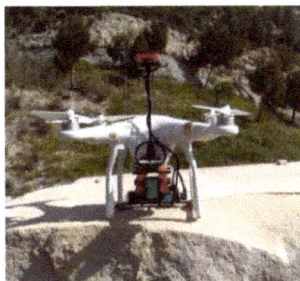

Figure 3. Phantom 3 DJI with the Parrot SEQUOIA camera and irradiance sensor, used for the aerial survey.

UAV imagery was captured at four different agricultural plots (Figure 1). These four plots were selected with expert criteria for their representativeness in terms of levels of damage and type of cultivated rice and were all near the potential harvesting date. After initial field inspection by rangers, Plots 1, 2, and 4 appeared to present damages causing loss of rice cover in many areas, showing a variety of spatial patterns (both scattered and aggregated). Initial inspections of Plot 3 did not detect damage to the rice crop. The four plots may have been treated with a variety of agricultural treatments in previous months, and plot 4 included a different rice variety. Images were captured during the same day in consecutive hours before and after the noon. The flying altitude was 80 m for three plots with an average GSD (Ground Sample Distance) of 8 cm and an average area of 12 ha. The flying altitude for the fourth flight was of 100 m with an average GSD of 10 cm. The flight speed was 5 m/second with 80% forward and sideways overlap. Images were captured on 17 August 2017, a few days before the harvest, when the maximum crop damages could be observed. Before each flight, images of the Sequoia calibration panel (by AIRINOV) were captured in order to apply radiometric calibration and corrections from reference reflectance maps. Crop monitoring requires accurate calibration [36], therefore, we ensured application of the same methodology in all the flights: the target is level with the ground and not at an angle; the target is not affected by shadows or reflections of eventual surrounding objects; and the panel images are taken around 1 meter away from the panel and not facing the sun. Several images were taken before the flight to ensure we obtained images with lighting conditions as similar as possible to the ones during the flight. At the moment of application of the radiometric calibration we chose the most appropriate panel picture for each flight.

Three of the plots were totally surveyed, corresponding to areas between 3 and 5 ha each, and the fourth plot had a total area of 194.60 ha, and a representative surface of 17 ha (8.7%) was surveyed.

3.3. UAV Imagery Processing

UAV data was processed into a multispectral orthomosaic with Pix4Dmapper Pro 4.2 following the "Ag Multispectral" template. Pix4D is based on the Structure from Motion algorithms and also integrates computer vision techniques with photogrammetric algorithms [37,38] to obtain high accuracy in aerial imagery processing [39,40]. Pix4Dmapper Pro computes key points in the single images and uses them to find matches between images. From these initial matches, the software runs several automatic aerial triangulation steps, bundle bloc adjustments, and camera self-calibration steps iteratively until optimal reconstruction is achieved [29]. Then, a densified point cloud is generated to obtain a highly detailed digital surface model (DSM) that will be used to generate the final orthomosaics and reflectance maps for every plot (Figure 4).

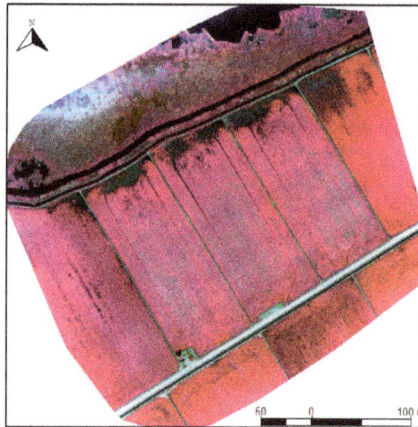

Figure 4. UAV color infrared composition: NIR (near-infrared) + RED + GREEN; for one of the agricultural plots (pixel resolution: 10 cm). Damages caused by Western Swamphen are shown in the northern part of the rice plots (dark colors correspond to water without rice plants). In the upper part of the image there is an important wetland, which is differentiated by a more dispersed wet vegetation zone and a fragment of water body (with dark-blue color). Surveyed plot number 1.

The reflectance maps were built by applying radiometric calibrations and corrections. First, radiometric calibration was performed by using the calibration target images that enable one to have an absolute reference of radiometry for that day, and finally makes it possible to compare data coming from several flights. Second, the radiometric corrections "Camera and Sun irradiance" aimed at correcting terrain reflectance was also applied. Within this, the Pix4Dmapper first uses the values of ISO, aperture, shutter speed, sensor response, and optical system and vignetting registered in the text metadata files (EXIF and XMP tags) for every single photogram to correct the camera. Then, the incoming sunlight irradiance is corrected with the information provided by the sun irradiance sensor. This sensor provides information on the light conditions during the flight within the same spectral bands as the one captured by the multispectral sensor and it is registered in the same text metadata files described above. This "Camera and Sun irradiance" correction normalizes the images captured during the flight and, thus, allows one to compare images taken in different illumination conditions. Pix4Dmapper applies this calibration and correction process to every single photogram just before it achieves the final reflectance orthomosaic for every spectral band (Figure 3). We have calculated the mean error in x-y coordinates by means of geolocation four-points-per-survey through screen digitalization in the UAV final orthomosaic and in the 1:1000 official orthophotos. We have calculated the difference between both points sets and the mean error in x and y is 0.56 m and 1 m, respectively.

We calculated the NDVI index for the UAV images. The normalized difference vegetation index NDVI [41] quantifies vegetation by measuring the difference between near-infrared (NIR), which vegetation strongly reflects, and red light (R), which vegetation absorbs with:

$$NDVI = \frac{NIR - R}{NIR + R} \qquad (2)$$

NDVI ranges from −1 to +1; negative values in wetland habitats are usually related to water surfaces, while positive values close to 0 represent low green vegetation cover and values close to 1 are usually associated with vegetation with dense green foliage. In this work, the NDVI index allowed us to identify no-greenery vegetation pixels, with or without water pixels, and also low greenery pixels, representing the damages caused by Western Swamphen in rice crops. It has been previously found with images from Sentinel-2 [42] that the use of vegetation indices improves the results of crop classifications over the use of the respective individual bands.

A simple Jenks classifier was applied separately for every flight and provided the highest discrimination ability to create a binary green vegetation/no-vegetation map (Figure 5). The Jenks classifier determined a class break value that produced two natural groupings by minimizing each class's average deviation from its mean, while maximizing each class's deviation from the other class's mean. The thresholds to separate the non-vegetation from green vegetation with NDVI data were different for each surveyed plot, taking values between 0.56 and 0.79, depending on the variety and the state of vegetation development of each plot.

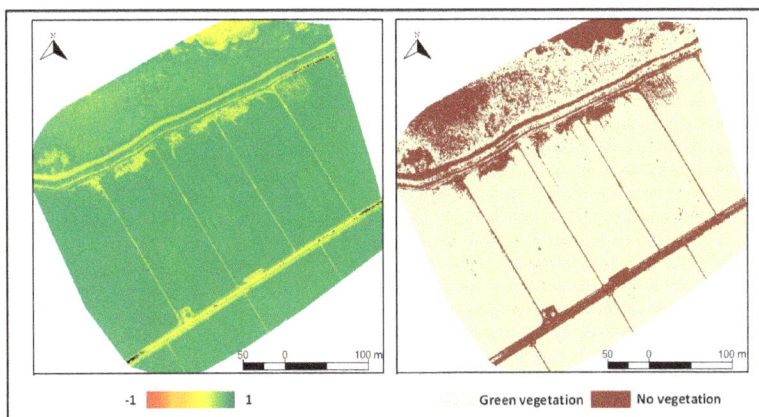

Figure 5. Surveyed plot 1: UAV-NDVI index on the left, and non-vegetation mask on the right. It is possible to appreciate an important lack of vegetation in the north part of the plot and some scattered damages in the middle and in the south part.

A 10 × 10 m mesh was created for each surveyed plot from the August 24 Sentinel-2 image pixels. The proportion of water and the proportion of non-vegetation for every 10 × 10 m square was calculated by obtaining a proportion of no-greenery vegetation for both indices (Figure 6). Values near 1 indicated a higher proportion of damage and values near 0 indicate lower or null proportion of damage. From now on we will refer to this 10 × 10 m information as UAV-crop damage.

3.4. Selecting Pure Rice Crops Squares

In order to reduce noise from the analyses, we selected the 10 × 10 m squares that completely contained rice crops (in contrast with other pixels that also contain other land uses, such as rural tracks) (Figure 6). Two sources of information were used to make this selection: (1) Agricultural Plot Map from the Catalan government, henceforthm SIGPAC (Spanish acronym of Geographical Information System of Agricultural Plots); and (2) all the available April–August 2017 Sentinel-2 cloud free images. SIGPAC is a free downloadable cartography provided by the government of Spain, which is renewed on a year basis. Pixels that were totally or mostly within a plot of "Arable Earth" category were selected. However, this cartography scale does not include all water channels. For this reason, Sentinel-2 imagery was also used to identify them. Since the water is not distributed equally throughout the year on all channels, we selected the maximum NDWI value for every pixel of the NDWI time series described in Section 3.1 to identify the maximum water that can flow at any time of the year. Finally, with the two sources we computed a water channel mask and we applied it to all UAV-crop damage and Sentinel2-NDVI data. In addition, it was necessary to filter some pixels manually that the described methodology had not recognized (especially in the sampled plots).

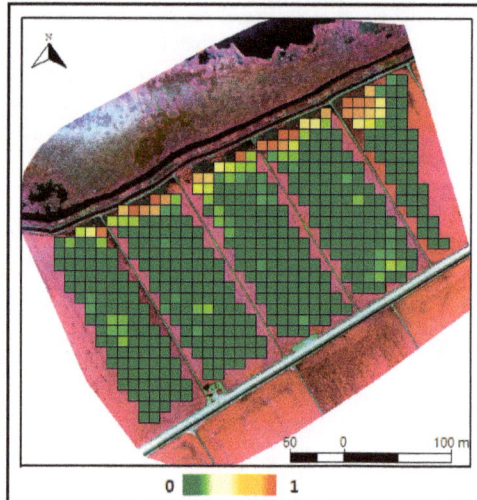

Figure 6. Selected 10 × 10 m cells (co-registered with Sentinel-2 pixels) for surveyed plot number 1. The image shows the non-vegetation ratio calculated with UAV images for every Sentinel-2 pixel used in the model calibration, from 0 to 1, where 0 represents green vegetation, or non-damage, and 1 non-vegetation proportion, or damaged. The background image corresponds to the UAV infrared composition image. It is possible to detect some important damages due to an important lack of vegetation in the north of the plot, and some pixels with a lower damage ratio due to scattered damage in the rest of the plot.

3.5. Calibrating Sentinel2 with UAV-Crop Damage Information

We calibrated Sentinel2-NDVI with UAV-crop damage information for the 10×10 m cells. We used a generalized linear mixed effect model (GLMM) that predicted the proportion of damage as a function of NDVI value derived from Sentinel, and incorporated each plot as the random factor. We adopted a logistic regression with a logit link function (Equation (3)).

$$ln\left(\frac{UAV_{damage_{ratio}}}{1 - UAV_{damage_{ratio}}}\right) = a + b * Sentinel2\_NDVI \qquad (3)$$

This logistic regression type was chosen because it allowed us to better fit values that cannot range below 0 or above 1. The mixed model regression is suitable to use when some of the observations share common sampling units that can gather some of the variance not directly explained by the fixed factors. In this case each of the agricultural plots may hold differences in rice variety and phenology, seed time, or irrigation conditions, but similarities in each of the cells within each plot. The variability of plot features is included through the random factor of the mixed model.

We checked for model residuals distribution through a fitted residual plot, and we evaluated model goodness-of-fit through the marginal R^2 (explained by fixed factors), conditional R^2 (explained by both fixed and random factors), and the Root Mean Square Error (RMSE). The model was validated by applying a cross-validation method using 80% of observations for calibration and 20% for testing. Predicted values for test observations were compared to observed values with a linear regression obtaining an adjusted r-squared and a test RMSE. To minimize the bias of the subsampling test data, we ran a 10-fold random subsampling cross-validation and we finally obtained an average adjusted r-squared and RMSE for the ten sub-validations.

3.6. Crop Damage Evaluation at the Plot Level: Translating the Final Models into Useful Products for Landscape Managers

The management unit for the landscape managers is the SIGPAC agricultural plot. The payments for damages are granted based on damage that occurred in each agricultural plot and quantified in hectares, so our goal was to provide a measure of damage per SIGPAC plot. We assume that the damages that cause large loss of vegetation, as the ones observed in the north of the surveyed plot number 1 (from Figure 4 to Figure 6), are the minority, and the majority of damages represent a small proportion of the 10 m Sentinel-2 pixel (check photographs in Figure 2). We also observe that most parts of 10 m pixels in the whole Delta have an estimated damage ratio under 1% that could not be considered as real damage caused by Wester Swamphen. Thus, we needed to identified pixels with significant crop damage that would be caused by the bird and then the final damage map would correspond to the sum of the full pixel surfaces, with significant damage values for each agricultural plot.

First, we needed to identify this significant crop damage threshold. To proceed with this identification, we tested different thresholds: 4%, 6%, 8%, 10%, and 20%. The selection of these thresholds took into account a compromise between level of significant damage (>1%) and number of damaged pixels selected within every threshold. Low thresholds selected a great number of pixels but may have included pixels that were not actually damaged by birds. Instead, high thresholds selected only pixels with large loss of vegetation, avoiding also damaged pixels without large loss of vegetation. Then, the sum of the damaged pixels for each threshold in each surveyed plot was calculated. Finally, this different estimations at the agricultural plot level were validated [30,43] with ground information from Rangers Service from the Catalan government, which follows an established protocol to assess crop damage in the field. Once a farmer has detected some damage caused by the Western Swamphen, he claims the assessment on the field by the Rangers Service to quantify damages and receive the corresponding payments. This assessment is performed on the ground, next to the affected plot, and it is only based on expert criteria. It does not constitute an exact quantification of the damages but it offers an approximate value to evaluate models' predictions.

4. Results

The four surveyed plots showed different ranges of UAV-crop damages and Sentinel2-NDVI at a 10 m pixel level. Plot number 1 and 2 held a wide UAV-crop damage range (0 to 0.97, and 0 to 0.93, respectively, Table 1), whereas the UAV-crop damage range in plot number 3 was between 0 and 0.36, and the crop UAV-crop damage for plot number 4 was always under 0.04. This indicated that plots 1 and 2 have larger parts with loss of vegetation, while the damage in crops 3 and 4 are smaller and dispersed throughout the plot. Sentinel2-NDVI range of values also showed clear differences: plot 1 had the widest range (0.32 to 0.78), plot number 2 ranged between 0.38 and 0.74, plot number 3 had Sentinel2-NDVI ranges between 0.38 and 0.95, and plot 4 showed a small range above 0.5, specifically between 0.52 and 0.78.

Table 1. Ranges of UAV-crop damage and Sentinel2-NDVI (Normalized difference vegetation index) per pixel for all the 10 m pixels in the surveyed plots.

Plots	UAV-Crop Damage Range	Sentinel2-NDVI Range
1	0–0.97	0.32–0.78
2	0–0.93	0.38–0.74
3	0–0.36	0.38–0.95
4	0–0.04	0.52–0.78

The selected fitted generalized linear mixed model between UAV-damage ratio and Sentinel2-NDVI achieved good model adjustments, with a RMSE of 0.054 (Table 2). The random effect of the variable "surveyed plot" appeared significant (<0.001), meaning that the intercept of the relationships between Sentinel2 and UAV-damage differed by plot. The model presented a goodness of fit explained by fixed effects (Sentinel2-NDVI) of 48%, described as marginal R^2 in Table 2, and explained by both fixed and random (surveyed plot) effects of 81%, described as conditional R^2 also in Table 2. The cross-validation decreased its prediction ability in comparison to the full calibrated model, but still achieved relevant performances, with an average of 57% in the regression between observed and predicted values and a RMSE of 0.102 (Table 2). The Sentinel2-NDVI index exerted a negative influence in predicting observed damage: an increase in Sentinel2-NDVI values decreases the proportion of Western Swamphen damage described by UAV (Figure 7).

Table 2. Goodness-of-fit estimate values for the fitted model and an average of 10 repetitions of model validation, with 20% of the dataset randomly selected for each repetition represented by the Marginal, Conditional and Adjusted coefficient of determination (R^2) and the Root Mean Square Error (RMSE).

Model Goodness-of-Fit Estimate			Model Validation (10-Fold 20% Dataset Validation)	
Marginal R^2	Conditional R^2	RMSE	Adjusted R^2	RMSE
0.481	0.814	0.054	0.572	0.102

The model predicted low values of damage proportion, close to 0, in most parts of the pixels with Sentinel2-NDVI values above 0.55 (blue line in Figure 7). NDVI values under 0.55 predicted damages ratios from 0.05 to 1 (corresponding to 5% and 100% respectively). Most pixels had a damage ratio under 0.20, decreasing considerably the number of pixels with damage ratios above 25%. However, in contrast to what was expected, there are some pixels with damages between 0.12 and mostly up to 0.25, with relatively high Sentinel-NDVI values, specifically with values between 0.6 and 0.8 (Figure 7), most of them belonging to plot 3.

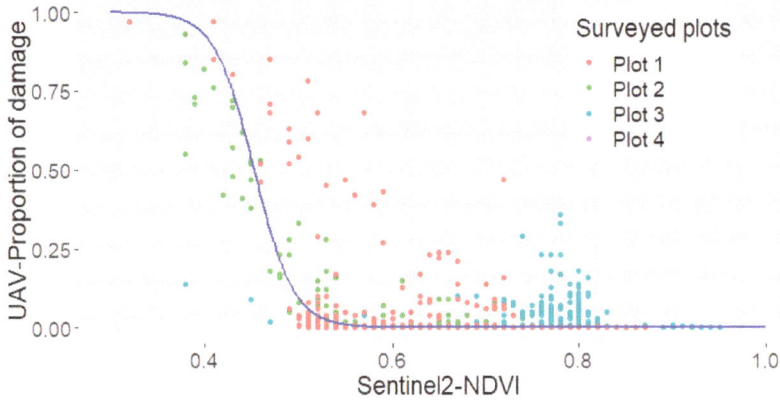

Figure 7. Scatterplot showing the relationships between UAV-crop damages and Sentinel2-NDVI. The blue line shows the best-fit model. Every surveyed plot is represented in different colors. Plot 4 (violet color) has got very low UAV-crop damages and high Sentinel2-NDVI values, so it is barely visible on the graph.

We applied the resulting model to the whole August 24 Sentinel-2 NDVI product for the Ebro Delta. The final map informed about proportion of damage per pixel with values between 0 and 1 (Figure 8). The predicted crop damage map identified the pixels with large areas of no-vegetation, with colors from yellow to red in Figure 8. Most part of the Delta had values close to 0, represented by green colors (Figure 8).

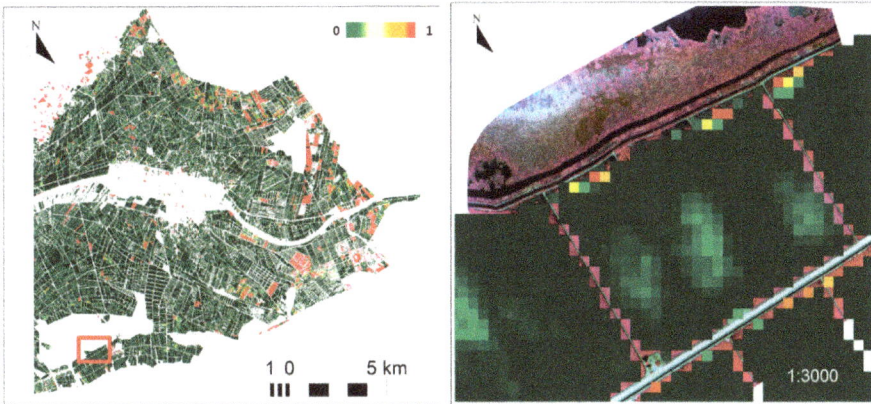

Figure 8. Predicted crop damage according to the NDVI model for the whole Delta (**left**) and a zoom to surveyed plot 1, in the south west of the Delta (**right**). Both figures take values from 0 to 1, colored from green to red, respectively, representing a pixel ratio damage from 0 to 1.

Finally, we applied the defined damage thresholds (4%, 6%, 8%, 10%, and 20%) to calculate the estimated agricultural plot damage (Figure 9). Once a pixel has a predicted damage above this threshold, the entire pixel is accounted as damaged. Pixels above these thresholds represented 13%, 12%, 11.6% 11.2%, and 10% of the total predicted surface, respectively. Therefore, damages ratios under 4% represent the majority of the Delta pixels (87% of total crop pixels).

The validation of the different thresholds with ground information from the Rangers Service showed an underestimation of damages in plots 1 and 2 (Table 3), even in the lowest damage ratio threshold. For plot 1, the largest threshold of damage assessed (20%) included some of the pixels with higher damage identified visually (Figure 9, northern part). On the contrary, and unexpectedly, the lowest threshold (4%) did not include all the high damage pixels identified visually, but it is also important to note that this threshold correctly identifies the scattered damages in the center of the plot. In plot 3, predicted damages are virtually null, similar to the information gathered by the Rangers Service (Table 3). Finally, there is a slight overestimation of predicted damages in plot 4, although as the damage threshold per pixel rises, the total value of the plot is closer to the quantification by the Rangers Service (Table 3), being the 20% threshold, which is slightly lower than Rangers Service quantification.

Table 3. Model validation at the plot level for the four surveyed plots with ground level Rangers Service information. This table shows the plot surface (ha) for each surveyed plot from the SIGPAC (Geographical Information System of Agricultural Plots from Catalan Government) plot surface information, the Rangers Service damage quantification, and predicted damage per agricultural plot at different thresholds, corresponding to different % of damage per pixel (4%, 6%, 8%, 10%, and 20%).

Plot Code	Plot Surface (ha)	Rangers Service (ha)	Predicted Damage per Agricultural Plot (ha) at Different Thresholds				
			>4%	>6%	>8%	>10%	>20%
1	4.53	1.09	0.47	0.29	0.16	0.11	0.06
2	5.16	0.66	0.52	0.47	0.45	0.42	0.37
3	3.37	No damage	0.01	0.01	0.01	0.01	0.01
4	194.60	2.19	2.51	2.43	2.36	2.35	2.18

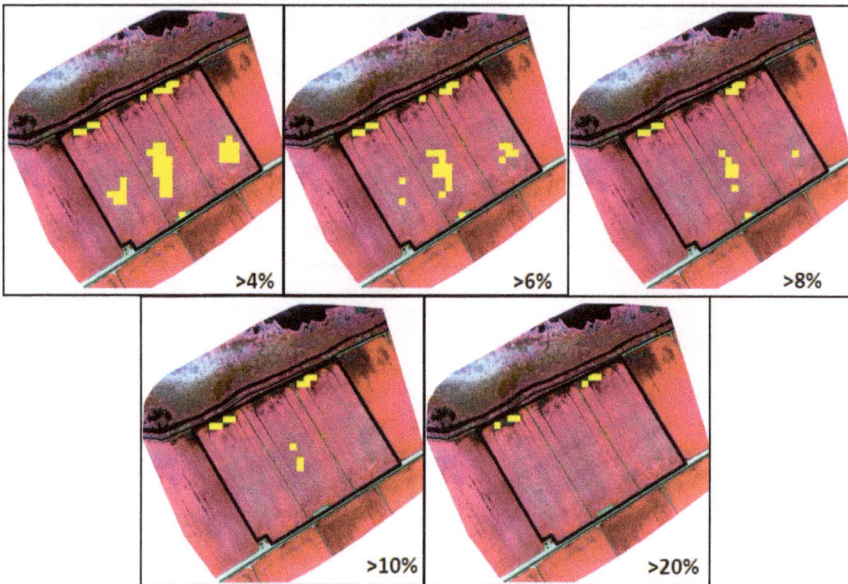

Figure 9. The yellow pixels show the selected damaged pixels at different thresholds of damages (4%, 6%, 8%, 10%, and 20%) in plot 1. There are 6 pixels with damage ratios above 20%, 11 pixels with damage ratios above 10%, 16 pixels with damage ratios above 8%, 29 pixels with damage ratios above 6%, and 47 pixels with damage ratios above 4%.

5. Discussion

The quantification of Western Swamphen crop damage is a big challenge due to its fine grained patterns. The methodology developed in this work, and especially the results and validation of the regression model, appear promising for quantifying this type of damage with Sentinel-2 imagery. The prediction of damages at the agricultural plot level with a previous filter of pure crop pixels is also promising. Previous analyses including pixels containing water channels, roads, pathways, or other land covers showed weaker relationships and major errors (not shown).

The model can more-easily predict damages that produce important loss of vegetation, and it has more difficulty in identifying the less-intense and scattered damages, i.e., less than 25% of damage per 10 m pixel. The scattered damages are able to be detected from UAV but are more difficult to appreciate at Sentinel-2 level because the NDVI values in these pixels remains high. The model presents important limitations to predict damages that only affect the rice grains without decreasing plant vigorousness, and even more if the damaged areas are covered by some type of vegetation, such as algae, which also produces high values of greenery. These types of damages are also arduous to detect with UAV multispectral imagery.

The Sentinel-2 10 m resolution resulted in consistent relationships with UAV information, since the fitted model achieved a good adjustment, although differences appeared among the four surveyed plots that may decrease the adjustment in the cross-validation. We assume that there are not important calibration differences between flights that could influence these results because we have ensured we applied the same methodology to capture calibration panels pictures and we also assume that the reflectivity is quite similar in the four flights due to the hour of the flights (few hours before and after the noon) and the date of the year (August 17th) in the Delta Ebro region. These differences might be related mainly to the differences in the pressure exerted by the species on the crop, rice variety, or agricultural treatments, or due to minor differences in phenology state, but these effects must be minimal because all fields were close to being harvested. These differences increase the uncertainty of the model, but the mixed generalized model is appropriate because it allows incorporation of such differences between crops as a random factor and derivation of one single model for the whole Delta area. The use of a single model for the whole Delta that incorporates the intrinsic variability of rice crops facilitates the transfer of the methodology to the landscape managers in order to use an objective methodology to quantify damages and define economic compensation strategies.

The Sentinel-2 10 m information resolution also allows a promising prediction of damages at the agricultural plot level. The selection of an adequate significant damage threshold shows good potential for identifying scattered damages at this scale. Considering a low damage ratio pixel as a whole affected pixel seems to counteract the difficulties of detecting scattered damages at the 10 m pixel level. This may be reasonable, because according to expert knowledge, scattered damages may show aggregated patterns, therefore, close to some of the detected damages there may be damages that effect rice production but produce low NDVI decrements that are very difficult to detect, even with UAV. The filtering of agricultural pure pixels may produce an underestimation of damages in pixels with important loss of vegetation because the majority of pixels with large loss of vegetation are near agricultural crop edges next to roads and water channels, and some of them they may have been excluded from the analysis, meaning they are not accounted for in the total damaged surface per agricultural plot. On the other hand, automatic pixel filtering also has other limitations in the other direction—there may also be an overestimation of damages in pixels with a small fraction of other uses that have not been filtered by the automatic methodology.

One of the highlights of this project has been the use of multispectral information. The NIR channel information from UAV allowed us to describe the characteristics of the vegetation, which are not well described by using the visible spectrum. At the same time, multispectral information enabled the generation of products more related to the multispectral medium resolution information derived from satellites as Sentinel-2 or Landsat. Previous works applying calibration of multispectral remote sensing measurements from UAV are very few and are aimed at quantifying other ecosystem attributes, such as fire severity [30,31]. In these cases, correlations between UAV and medium resolution remote sensing imagery have been found to be higher than the ones presented in this work. This may be because the damages caused by large forest fires are more apparent and visually-identifiable from medium resolution remote sensors than the Western Swamphen damages.

This calibration can be used to estimate damages in Sentinel images in other temporal or spatial contexts. Further validations are fostered in order to better test the capability to properly predict these kinds of damages. However, it is important to note that there may be other animal species that produce similar damages as the ones produced by the Western Swamphen. Moreover, images might detect other agronomic variations of the crops themselves due to diseases, germination problems, environmental variations, etc., with similar responses to the damage produced by animals. For this reason, a complementary expert identification of the plots that have been damaged by this species will generally be required in order to correctly estimate the final cause of damage to the rice crop.

Finally, we conclude that the models in this work improve the cost-efficiency of the quantification of Western Swamphen damage in large rice crop areas, especially in terms of objectivity. However, the methodology presents some weakness that should be addressed in the future to enhance the robustness of the models. We believe that the main points to consider could be the following:

- Exploring other UAV vegetation indices that could improve the regressions with Sentinel-2 imagery in order to better discriminate scattered damages, and also identify damaged spots that have been covered with algae or other vegetation and damages that only affect the rice grains.
- Analyzing in greater depth the potential relationship between different rice types and the detectable damage. Different varieties of rice are cultivated, each one with different morphology, and these differences could affect the detectability of the damage.
- Analyzing the effects, including the information from neighboring pixels.
- Improved mapping of water channels, roads, and pathways cartography in order to extract easily the Sentinel-2 pixels with these land covers.
- Working on a reliability map aimed at offering the land manager proper information about the real quality of the final model at the plot level.

6. Conclusions

We have introduced here a new methodology to use Sentinel-2-based NDVI as a proxy to quantify rice crop plot damages caused by Western Swamphen. Furthermore, we have shown how satellite-derived crop damage estimates can be calibrated by using UAV multispectral imagery onboard a low-cost unmanned vehicle. Our approach allows estimation of rice crop damage at large scales with relatively good accuracy. This approach offers a good, cost-effective alternative to high resolution imagery from more expensive, finer-resolution commercial satellites. Following works on knowledge transfer will improve the preparation of quality maps that could allow assessment with a greater accuracy of crop damages. Providing information about the quality of the products and their limitations is crucial for a useful knowledge transfer process.

Author Contributions: Conceptualization, G.B. and L.B. Formal analysis, M.P. Funding acquisition, R.G. Investigation, M.P. and L.B. Methodology, M.P. Project administration, G.B., R.G., and L.B. Resources, J.B. Supervision, L.B. Validation, A.C. Writing—review and editing, M.P., A.D., and L.B.

Funding: This research was funded by Ministry of Territory and Sustainability, Government of Catalonia.

Acknowledgments: We thank the Parrot Innovation Grant for the provision of the multispectral camera and the use of the Pix4D software. We also thank the Ebro Delta Natural Park and the Rangers Service of the Catalan Government for the facilities offered during the capture of UAV images.

Conflicts of Interest: The authors declare no conflict of interest.

References

1. Barua, M.; Bhagwat, S.A.; Jadhav, S. The hidden dimensions of human-wildlife conflict: Health impacts, opportunity and transaction costs. *Biol. Conserv.* **2013**, *157*, 309–316. [CrossRef]
2. Redpath, S.M.; Young, J.; Evely, A.; Adams, W.M.; Sutherland, W.J.; Whitehouse, A.; Amar, A.; Lambert, R.A.; Linnell, J.D.C.; Watt, A.; Gutiérrez, R.J. Understanding and managing conservation conflicts. *Trends Ecol. Evol.* **2013**, *28*, 100–109. [CrossRef]
3. Moreno-Opo, R.; Pique, J. Reconciling the conservation of the purple swamphen (Porphyrio porphyrio) and its damage in Mediterranean rice fields through sustainable non-lethal techniques. *PeerJ* **2018**, 1–19. [CrossRef] [PubMed]
4. Karanth, K.K.; Gopalaswamy, A.M.; Prasad, P.K.; Dasgupta, S. Patterns of human-wildlife conflicts and compensation: Insights from Western Ghats protected areas. *Biol. Conserv.* **2013**, *166*, 175–185. [CrossRef]
5. Agarwala, M.; Kumar, S.; Treves, A.; Naughton-Treves, L. Paying for wolves in Solapur, India and Wisconsin, USA: Comparing compensation rules and practice to understand the goals and politics of wolf conservation. *Biol. Conserv.* **2010**, *143*, 2945–2955. [CrossRef]
6. Villero, D.; Pla, M.; Camps, D.; Ruiz-Olmo, J.; Brotons, L. Integrating species distribution modelling into decision-making to inform conservation actions. *Biodivers. Conserv.* **2017**, *26*, 251–271. [CrossRef]
7. López-Bao, J.V.; Frank, J.; Svensson, L.; Åkesson, M.; Langefors, Å. Building public trust in compensation programs through accuracy assessments of damage verification protocols. *Biol. Conserv.* **2017**, *213*, 36–41. [CrossRef]
8. Lane, S.J.; Azuma, A.; Higuchi, H. Wildfowl damage to agriculture in Japan. *Agric. Ecosyst. Environ.* **1998**, *70*, 69–77. [CrossRef]
9. Pernollet, C.A.; Simpson, D.; Gauthier-Clerc, M.; Guillemain, M. Rice and duck, a good combination? Identifying the incentives and triggers for joint rice farming and wild duck conservation. *Agric. Ecosyst. Environ.* **2015**, *214*, 118–132. [CrossRef]
10. IUCN Redlist; Birdlife International. *Gavia Stellata: European Red List of Birds*; Birdlife International: Cambridge, UK, 2015; ISBN 9789279474507.
11. Tucker, M.; Heath, M.F.; Tomialojc, L.; Grimmett, R. *Birds in Europe: Their Conservation Status*; Birdlife Conservation: Cambridge, UK, 1994; ISBN 10: 0946888299/ISBN 13: 9780946888290.
12. Kwak, Y.; Shrestha, B.B.; Yorozuya, A.; Sawano, H. Rapid Damage Assessment of Rice Crop after Large-Scale Flood in the Cambodian Floodplain Using Temporal Spatial Data. *IEEE J. Sel. Top. Appl. Earth Obs. Remote Sens.* **2015**, *8*, 3700–3709. [CrossRef]
13. Möller, M.; Gerstmann, H.; Gao, F.; Dahms, T.C.; Förster, M. Coupling of phenological information and simulated vegetation index time series: Limitations and potentials for the assessment and monitoring of soil erosion risk. *Catena* **2017**, *150*, 192–205. [CrossRef]
14. Navrozidis, I.; Alexandridis, T.K.; Dimitrakos, A.; Lagopodi, A.L.; Moshou, D.; Zalidis, G. Identification of purple spot disease on asparagus crops across spatial and spectral scales. *Comput. Electron. Agric.* **2018**, *148*, 322–329. [CrossRef]
15. Inglada, J.; Arias, M.; Tardy, B.; Hagolle, O.; Valero, S.; Morin, D.; Dedieu, G.; Sepulcre, G.; Bontemps, S.; Defourny, P.; Koetz, B. Assessment of an operational system for crop type map production using high temporal and spatial resolution satellite optical imagery. *Remote Sens.* **2015**, *7*, 12356–12379. [CrossRef]
16. Belgiu, M.; Csillik, O. Sentinel-2 cropland mapping using pixel-based and object-based time-weighted dynamic time warping analysis. *Remote Sens. Environ.* **2018**, *204*, 509–523. [CrossRef]
17. Johansen, K.; Sallam, N.; Robson, A.; Samson, P.; Chandler, K.; Derby, L.; Eaton, A.; Jennings, J. Using GeoEye-1 Imagery for Multi-Temporal Object-Based Detection of Canegrub Damage in Sugarcane Fields in Queensland, Australia. *GIScience Remote Sens.* **2018**, *55*, 285–305. [CrossRef]
18. Ćulibrk, D.; Lugonja, P.; Minić, V.; Crnojević, V. Water-stressed crops detection using multispectral worldview-2 satellite imagery. *Int. J. Artif. Intell.* **2012**, *9*, 123–139. [CrossRef]
19. Zhao, J.L.; Zhang, D.Y.; Luo, J.H.; Huang, S.L.; Dong, Y.Y.; Huang, W.J. Detection and mapping of hail damage to corn using domestic remotely sensed data in China. *Austral. J. Crop Sci.* **2012**, *6*, 101–108.

20. Anderson, K.; Gaston, K.J. Lightweight unmanned aerial vehicles will revolutionize spatial ecology. *Front. Ecol. Environ.* **2013**, *11*, 138–146. [CrossRef]
21. Zhang, J.; Hu, J.; Lian, J.; Fan, Z.; Ouyang, X.; Ye, W. Seeing the forest from drones: Testing the potential of lightweight drones as a tool for long-term forest monitoring. *Biol. Conserv.* **2016**, *198*, 60–69. [CrossRef]
22. Ribeiro-Gomes, K.; Hernandez-Lopez, D.; Ballesteros, R.; Moreno, M.A. Approximate georeferencing and automatic blurred image detection to reduce the costs of UAV use in environmental and agricultural applications. *Biosyst. Eng.* **2016**, *151*, 308–327. [CrossRef]
23. Huang, Y.; Reddy, K.N.; Fletcher, R.S.; Pennington, D. UAV Low-Altitude Remote Sensing for Precision Weed Management. *Weed Technol.* **2018**, *32*, 2–6. [CrossRef]
24. Zhang, C.; Kovacs, J.M. The application of small unmanned aerial systems for precision agriculture: A review. *Precis. Agric.* **2012**, *13*, 693–712. [CrossRef]
25. Stanton, C.; Starek, M.J.; Elliott, N.; Brewer, M.; Maeda, M.M.; Chu, T. Unmanned aircraft system-derived crop height and normalized difference vegetation index metrics for sorghum yield and aphid stress assessment. *J. Appl. Remote Sens.* **2017**, *11*, 026035. [CrossRef]
26. Su, B.F.; Xue, J.R.; Xie, C.Y.; Fang, Y.L.; Song, Y.Y.; Fuentes, S. Digital surface model applied to unmanned aerial vehicle based photogrammetry to assess potential biotic or abiotic effects on grapevine canopies. *Int. J. Agric. Biol. Eng.* **2016**, *9*, 119–130. [CrossRef]
27. Zhou, J.; Pavek, M.J.; Shelton, S.C.; Holden, Z.J.; Sankaran, S. Aerial multispectral imaging for crop hail damage assessment in potato. *Comput. Electron. Agric.* **2016**, *127*, 406–412. [CrossRef]
28. Su, J.; Liu, C.; Coombes, M.; Hu, X.; Wang, C.; Xu, X.; Li, Q.; Guo, L.; Chen, W. Wheat yellow rust monitoring by learning from multispectral UAV aerial imagery. *Comput. Electron. Agric.* **2018**, *155*, 157–166. [CrossRef]
29. Fernández-Guisuraga, J.M.; Sanz-Ablanedo, E.; Suárez-Seoane, S.; Calvo, L. Using unmanned aerial vehicles in postfire vegetation survey campaigns through large and heterogeneous areas: Opportunities and challenges. *Sensors* **2018**, *18*, 586. [CrossRef]
30. Fraser, R.H.; Van Der Sluijs, J.; Hall, R.J. Calibrating Satellite-Based Indices of Burn Severity from UAV-Derived Metrics of a Burned Boreal Forest in NWT, Canada. *Remote Sens.* **2017**, *9*, 279. [CrossRef]
31. Pla, M.; Duane, A.; Brotons, L. Potencial de las imágenes UAV como datos de verdad terreno para la clasificación de la severidad de quema de imágenes landsat: Aproximaciones a un producto útil para la gestión post incendio. *Rev. Teledetec.* **2017**, *2017*, 91–102. [CrossRef]
32. Bertolero, A.; Rivaes, S.; Mougeot, F.; Sánchez-Barbudo, I.S.; Andree, K.B.; Ibáñez, C. Sexing and Ageing the Purple Swamphen Porphyrio porphyrio porphyrio by Plumage and Biometry. *Ardeola* **2016**, *63*, 261–277. [CrossRef]
33. BOE. *Ley 42/2007, de 13 de diciembre, del Patrimonio Natural y de la Biodiversidad*; BOE: Madrid, Spain, 2007.
34. Vuolo, F.; Żółtak, M.; Pipitone, C.; Zappa, L.; Wenng, H.; Immitzer, M.; Weiss, M.; Baret, F.; Atzberger, C. Data Service Platform for Sentinel-2 Surface Reflectance and Value-Added Products: System Use and Examples. *Remote Sens.* **2016**, *8*, 938. [CrossRef]
35. Banu, T.P.; Borlea, G.F.; Banu, C. The Use of Drones in Forestry. *J. Environ. Sci. Eng. B* **2016**. [CrossRef]
36. Iqbal, F.; Lucieer, A.; Barry, K. Simplified radiometric calibration for UAS- mounted multispectral sensor. *Eur. J. Remote Sens.* **2018**, *51*, 301–313. [CrossRef]
37. Puliti, S.; Olerka, H.; Gobakken, T.; Næsset, E. Inventory of Small Forest Areas Using an Unmanned Aerial System. *Remote Sens.* **2015**, *7*, 9632–9654. [CrossRef]
38. Näsi, R.; Honkavaara, E.; Lyytikäinen-Saarenmaa, P.; Blomqvist, M.; Litkey, P.; Hakala, T.; Viljanen, N.; Kantola, T.; Tanhuanpää, T.; Holopainen, M. Using UAV-based photogrammetry and hyperspectral imaging for mapping bark beetle damage at tree-level. *Remote Sens.* **2015**, *7*, 15467–15493. [CrossRef]
39. Ruzgiene, B.; Berteška, T.; Gečyte, S.; Jakubauskiene, E.; Aksamitauskas, V.Č. The surface modelling based on UAV Photogrammetry and qualitative estimation. *Measurement* **2015**, *73*, 619–627. [CrossRef]
40. Zahawi, R.A.; Dandois, J.P.; Holl, K.D.; Nadwodny, D.; Reid, J.L.; Ellis, E.C. Using lightweight unmanned aerial vehicles to monitor tropical forest recovery. *Biol. Conserv.* **2015**, *186*, 287–295. [CrossRef]
41. Rouse, J.W.; Hass, R.H.; Schell, J.A.; Deering, D.W. Monitoring vegetation systems in the great plains with ERTS. *Third Earth Resour. Technol. Satell. Symp.* **1973**, *1*, 309–317.

42. Zhang, T.; Chen, W. Potential Bands of Sentinel-2A Satellite for Classification Problems in Precision Agriculture. *Int. J. Autom. Comput.* **2019**. [CrossRef]
43. Piñeiro, G.; Perelman, S.; Guerschman, J.P.; Paruelo, J.M. How to evaluate models: Observed vs. predicted or predicted vs. observed? *Ecol. Modell.* **2008**, *216*, 316–322. [CrossRef]

drones

MDPI

Article

A UAV-Based Sensor System for Measuring Land Surface Albedo: Tested over a Boreal Peatland Ecosystem

Francis Canisius [1,*], Shusen Wang [1,*], Holly Croft [2], Sylvain G. Leblanc [1], Hazen A. J. Russell [3], Jing Chen [2] and Rong Wang [2]

[1] Canada Center for Remote Sensing, Natural Resources Canada, Ottawa, ON K1S 5K2, Canada;
     sylvain.leblanc@canada.ca
[2] University of Toronto, Department of Geography, Toronto, ON M5S 3G3, Canada;
     holly.croft@utoronto.ca (H.C.); jing.chen@utoronto.ca (J.C.); wangr@geog.utoronto.ca (R.W.)
[3] Geology Survey of Canada, Natural Resources Canada, Ottawa, ON K1A 0E8, Canada;
     hazen.russell@canada.ca
*   Correspondence: francis.canisius@canada.ca (F.C.); shusen.wang@canada.ca (S.W.)

Received: 22 February 2019; Accepted: 12 March 2019; Published: 16 March 2019

Abstract: A multiple sensor payload for a multi-rotor based UAV platform was developed and tested for measuring land surface albedo and spectral measurements at user-defined spatial, temporal, and spectral resolutions. The system includes a Matrice 600 UAV with an RGB camera and a set of four downward pointing radiation sensors including a pyranometer, quantum sensor, and VIS and NIR spectrometers, measuring surface reflected radiation. A companion ground unit consisting of a second set of identical sensors simultaneously measure downwelling radiation. The reflected and downwelling radiation measured by the four sensors are used for calculating albedo for the total shortwave broadband, visible band and any narrowband at a 1.5 nm spectral resolution within the range of 350–1100 nm. The UAV-derived albedo was compared with those derived from Landsat 8 and Sentinel-2 satellite observations. Results show the agreement between total shortwave albedo from UAV pyranometer and Landsat 8 ($R^2 = 0.73$) and Sentinel-2 ($R^2 = 0.68$). Further, total shortwave albedo was estimated from spectral measurements and compared with the satellite-derived albedo. This UAV-based sensor system promises to provide high-resolution multi-sensors data acquisition. It also provides maximal flexibility for data collection at low cost with minimal atmosphere influence, minimal site disturbance, flexibility in measurement planning, and ease of access to study sites (e.g., wetlands) in contrast with traditional data collection methods.

Keywords: UAV; albedo; hyperspectral; Landsat 8; Sentinel-2

1. Introduction

Surface albedo, the fraction of the incident solar radiation that the surface reflects, controls the radiation absorption and microclimate conditions of soil and vegetation canopies, which affect physical, physiological, and biogeochemical processes such as evapotranspiration and ecosystem carbon cycle [1,2]. Albedo of vegetated land surfaces vary largely because of the seasonal changes of the land surface conditions and persistence changes of land use and land cover. Plant phenological cycles, including leaf growth and leaf fall, and temporal variations in foliar water and chlorophyll content control leaf optical characteristics modify the seasonal distribution of surface albedo [3–6]. Further, seasonal variations in snow cover cause large variations in surface albedo [7]. Clear-sky days tend to have lower surface albedo values than cloudy days in winter [8]. These changes can result in dynamic effects in surface albedo and ultimately ecological processes.

There are several operational surface albedo products from satellite remote sensing with spatial resolution ranging from approximately 0.5–25 km and temporal frequencies of daily to monthly [9]. Most notably these products include Moderate Resolution Imaging Spectroradiometer [10], Advanced Very High Resolution Radiometer [11], Polarization and Directionality of the Earth Reflectance [12] and Meteosat [13]. However, coarse resolution products are not detailed enough for studying ecological processes in small dynamic ecosystems such as wetlands. Retrieving surface albedo from high spatial resolution remotely sensed data (e.g., Landsat 8 and Sentinel-2) is methodologically challenging, in part because of the narrow angular sampling and the incomplete spectral sampling from a limited number of wavebands [2] and demands in-situ calibration and validation datasets. In-situ albedo measurements have long been using sensors mounted on research towers. These measurements, though, demand a high logistical requirement and only a few surfaces can be characterized and measured [2]. In order to account for a wide range of surface cover types, aerial based sensor system can play an important role in measuring surface albedo. Measurements from unmanned aerial platforms effectively bridges the observational scale between point-based tower measurements and satellite imageries [14].

Unmanned aircraft systems (UAS) are small planes capable of carrying small-sized sensors [15] and [14,16] use UAS mounted spectrometers or cameras to measure reflectance, which is used to calculate surface albedo or compare to MODIS products. The deployment of multi-rotor based UAV systems for remote sensing applications has been made possible by the miniaturization of technology, including accurate GPS systems, accelerometers, flight control systems and very light weight sensors [17]. To-date the emerging UAV technology has had only a few studies using UAVs to estimate surface albedo [14,16,18]. This paper describes multi-rotor based UAV system integrated with multiple small sensors, which are available in the market. The relatively low-cost UAV system carries multiple sensors that provide land surface parameters such as land cover, vegetation structure, total shortwave albedo, visible albedo, hyperspectral reflectance and vegetation indices at a particular time in different spatial sampling. System can acquire measurements over an area with multiple measurements that are comparable to the pixel size of high-resolution remote sensing (Landsat 8 and Sentinel-2) derived products. In order to demonstrate its capabilities, test albedo measurements were made in a protected wetland region where access is limited to study the local-scale variations in tandem with a ground based upward looking system and compared with the satellite derived albedo estimates of Landsat 8 and Sentinel-2. The comparison provides the necessary data for understanding the consistency between sensor measurements and satellite based estimates. This UAV system can help to recommend the in-situ albedo measurement protocol, which includes environmental conditions, viewing geometry, illumination geometry, properties of the target, measurement timing, instrument calibration and experimental design, for different land covers.

2. System Development

2.1. UAV Platform and Instruments

The UAV chosen for this system is the Matrice 600 (M600), DJI's new flying platform [19] designed for professional aerial photography and industrial applications (Figure 1). The Matric 600 was chosen as it has a number of mission critical features that are favorable for our research requirements. It has six rotors providing redundancy in case of rotor failure and enhanced stability. The UAV can recover itself from failure of a single battery because of its smart battery management system. The Matrice 600 is comparatively larger than many other drones; however, it is built of light-weight stiff carbon fibre and it is able to fold up for transportation. It has an approximately 40 minute hover time without a payload and 18 minutes with maximum payload (5.5kg). There is sufficient room for mounting of sensors. It has retractable landing gear, which ensures that the legs do not disturb sensor measurements. The Matrice 600 has a powerful operating system, which optimized flight control performance, operated by revolutionary A3 flight controller [19]. The flight control system has a transmitter and a mobile device, such as a tablet or smartphone, can be used to design flight paths and operate the drone. The Matrice 600 supports the DJI Go, the DJI Assistant 2 and the Litchi apps that give users a built-in flight simulator, HD view, battery level status, redundancy status, transmission strength, etc. Data regarding the use and operation of the UAV, flight telemetry data such as speed, altitude, compass, pitch, roll, battery life and information about the gimbal and camera and operation records are recorded in the system.

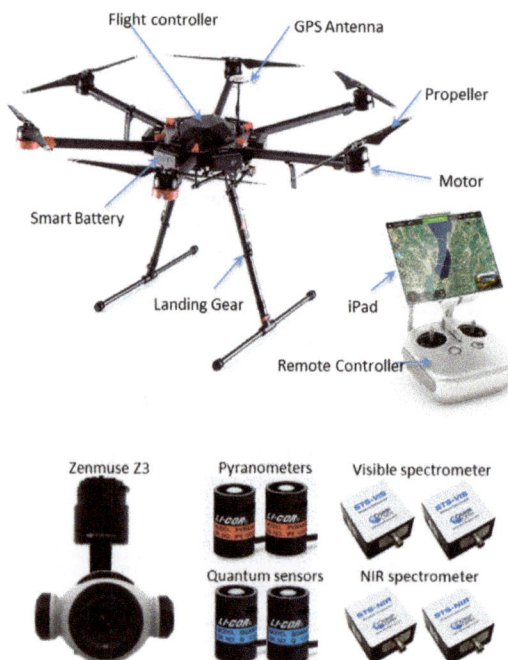

Figure 1. The Matrice 600 UAV, Zenmuse Z3 camera, pyranometers, Quantum sensors, and Visible and NIR spectraometers.

Instruments that were used in the system include a digital camera, a pair of broadband pyranometers, a pair of broadband quantum sensors, a pair of narrowband visible (VIS) spectrometers and a pair of narrowband near infra red (NIR) spectrometers (Figure 1). The Zenmuse Z3 digital camera supports 4K video recording (equivalent of 8.9 MP) at 30 frames per second and 12MP still photographs every 3 seconds. For high along track overlap, the 4K video is preferable as it has similar ground pixel resolution as the still photographs, but with a much high frequency. Zenmuse Z3 supports Micro SD cards with a capacity of up to 64GB to store high-resolution video data, photos and flight telemetry data.

The LI-200R Pyrometer (Li-Cor Inc., Lincoln, Nebraska USA, www.licor.com) measures global solar radiation (direct and diffuse) in the 400 nm to 1100 nm broadband range. It measures solar irradiance (the radiant flux incident on a receiving surface from all direction) received on a horizontal surface with a silicon photodiode mounted under a cosine-corrected acrylic diffuser [20]. The sensor output is a current (μA) signal that is directly proportional to hemispherical solar radiation (Wm^{-2}). The LI-190R Quantum sensor (Li-Cor Inc., Lincoln, Nebraska USA, www.licor.com) measures Photosynthetically Active Radiation (PAR) in μmol of photons m^{-2}s^{-1} [20]. It measures PAR between 400 nm to 700 nm broadband range, which vegetation uses for photosynthesis [20]. The sensors are connected to data loggers, which are programmed to collect data every 3 s. A calibration constant, which can be found in certificate of calibration, is used to convert the current signal into units of radiation (Wm^{-2}).

The spectral response functions of the LI-200R pyranometer and LI-190R quantum sensor in relation to Landsat 8 and Sentinel-2 satellite sensors are shown in Figure 2. The spectral response functions for the spectral bands sampled by the satellite sensors show a broad similarity in both position and bandwidth. The spectral response function of the quantum sensor is wide enough to cover the blue green and red spectral bands sampled in the visible range by the satellite sensors. The large difference, however, is between the satellite sensors and the pyranometer, and whilst the broadband pyranometer samples across the entire VIS-NIR spectrum, the contributions are weighted towards the NIR, with relatively less signals coming from shorter visible wavelengths, particularly in the blue-green region.

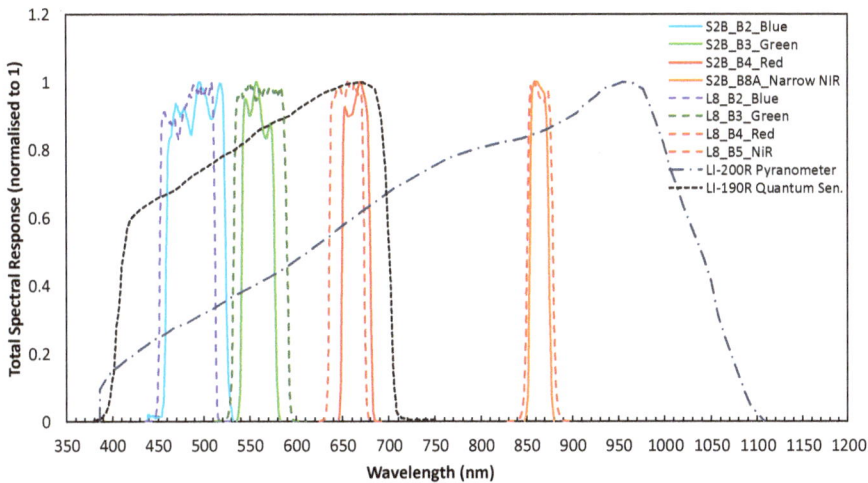

Figure 2. Relative spectral response functions (normalized to one) of LI-200R pyranometer, LI-190R Quantum sensor, Landsat 8 OLI and Sentinel-2 MultiSpectral Instrument (MSI) in visible and near infrared (VIS-NIR) range.

The STS-VIS Spectrometer (Ocean Optics, Largo, FL, USA, www.oceanoptics.com) measures the radiation within the visible range of 350 nm to 800 nm and the STS-NIR Spectrometer (Ocean Optics, Largo, FL, USA, www.oceanoptics.com) measures the radiation in the infrared region of 650 nm to 1100 nm with 1.5 nm spectral resolution. It is recommended to "warm up" the spectrometers prior to use for spectral measurement collection because the VIS and NIR spectrometers' arrays warm up at different rates that can cause spectral steps in the overlap region between VIS and NIR [21]. The spectrometers are connected to STS Developer's Kit [22] that contains a Raspberry Pi B+ microcomputer, a Wi-Fi™ dongle, a SD card containing all the software, a clock and a Lithium-Ion battery. The battery is able to provide 5 Volts and at least 1000 mA through a micro-USB connection. A laptop is used to communicate with the spectrometers and to setup the spectrometer parameters. OceanView allows complete control of setting the parameters for all system functions, such as: acquiring data, electrical dark-signal correction, boxcar pixel smoothing, and signal averaging. Integration time is an important parameter and it is recommended to adjust the integration time to acquire the maximum amount of light up to 85% of the spectrometer's capability [23]. Other advanced features support several data-collection options such as independently store and retrieve spectral data as ASCII files to disk using auto-incremented filenames.

2.2. *Systems Set-Up*

The system is composed of a UAV sub-system (Figure 3a) and a ground sub-system. The UAV sub-system includes a camera and four radiations sensors (Figure 3): one from each pair mentioned above. The camera is mounted using DJI 3-axis gimbal system onto the Matrice 600 and stream live HD video to the DJI or Litchi apps. One pyranometer and one quantum sensor are connected with a CR300 data logger and mounted on the UAV. Similarly, one VIS spectrometer and one NIR spectrometer are connected to a microcomputer and mounted on the UAV. The data logger and the microcomputer are powered by a separate Lithium-Ion battery. All the sensors are mounted on the UAV using a light-firm aluminum frame and are pointing down for measuring reflected radiation from the land surface.

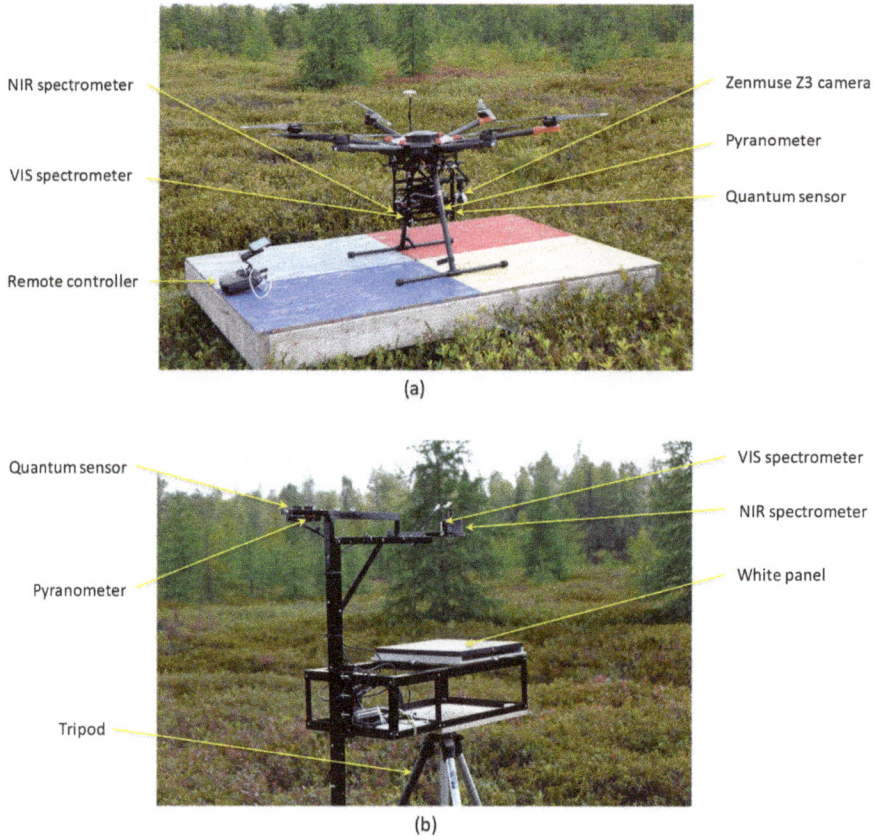

(a)

(b)

Figure 3. UAV (**a**) and ground (**b**) systems.

In addition to flight control system, the ground unit includes the second set of pyranometer and quantum sensor that are connected with a CR300 data logger and the second set of VIS and NIR spectrometers that are connected to a microcomputer. The ground pyranometer and the quantum sensor are pointed vertically upward to record downwelling solar radiation (direct radiation from the sun and diffuse radiation from the sky). The VIS and NIR spectrometers are set to face the calibrated white panel (~100% reference standard) to measure downwelling radiation. The ground unit is mounted on a tripod that helps to precisely level the system with a clear view of the sky or surface to ensure accurate measurements. Because the downwelling and reflected radiation measured by the sensors are used to estimate the surface albedo, it is recommended to operate the two system close to each other in order to minimize the differences between their respective ground footprints. The laptop is used to communicate with the UAV and ground sensors, and to set up clock time and the spectrometer parameters, such as integration time and acquisition timing before a flight. The spectrometers' microcomputer can be communicated with the laptop via wireless network to start, stop, and view the spectrometer data in real time.

2.3. Data Processing Chain

The data is processed using the workflow described in Figure 4. Data regarding the operation of the UAV including time, coordinates, speed, altitude, compass, pitch, roll, and information about the gimbal and camera are recorded in the system mission path file. Multispectral camera, ground and UAV data loggers (pyranometers and quantum sensors are connected) and ground and UAV microcomputers (VIS and NIR spectrometers are connected) write a millisecond level timestamp (from turn on time) and a real-world timestamp, based on their system's time, in their data files. Therefore, it is very important to synchronize the internal clocks to an accuracy of 1 second on all devices such as, the iPad, laptop, ground and UAV data loggers, and ground and UAV microcomputers. The timestamp is the common field (link) of all the datasets from the camera and the sensors, and is used to get the coordinates of the photos and sensor readings. The downward sensors are aligned within the airframe so that they are level when the UAV is in stable flight. Though the M600 is comparatively a stable platform, having stable flight is not always possible as the UAV platform orientation varies due to wind, its own rotation, and instability during acceleration and deceleration. Errors due to instrument tilting can be limited by omitting radiation data recorded when the pitch or roll of the UAV exceeded 3° [14].

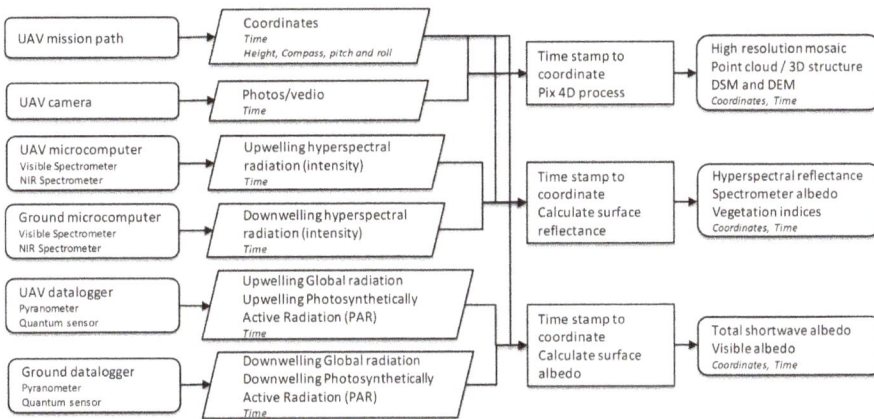

Figure 4. The data processing workflow.

The 4K videos are imported into Pix4D Mapper (www.pix4d.com) and one frame per second is extracted. The flight path information is used for georeferencing the extracted frames. This can be done without GPS-surveyed ground control points because the software pre-aligns the images based on the image matching technique to achieve accurate image alignment. The technique used by Pix4D to obtain a point cloud is a combination of Structure from Motion and photogrammetry [24]. These two techniques combined allow the retrieval of camera and target positions in a three dimensional system for all photographs. Geographic information can be added to the 3D scenes with ground control point and/or photographs with geo-information. The initial step creates a set of 3D matching points and the second steps densified that point cloud. A final step creates a high-resolution orthomosaics used to identify detailed land cover which will be used for the interpretation of spectral data and understanding the BRDF effects. Orthomosaics only create maps with 2D information, whereas point cloud gives 3D information such as DSM, DEM and structure of vegetation.

Spectrometer data is acquired in raw format as intensity value that depends on the integration time. First, the ground sensor measurements are adjusted with the calibration coefficients of the white panel for each wavelength. Dark noise (measured with the lens cap on) is subtracted from both intensity readings at each wavelength. The data from the UAV and ground sensors need to be aligned based on time/coordinate prior to calculation of reflectance. Reflectance of ground targets is calculated as stated in [25].

$$\rho'(\lambda) = \frac{I(S, \lambda)}{I(R, \lambda)} \tag{1}$$

where ρ' is nominal surface reflectance, λ is wavelength, I is the intensity recorded by the spectrometer for target surfaces (S) and the intensity adjusted with the calibration coefficients of the white reference (R), respectively. From the reflectance, different vegetation indices can be calculated.

The final stage in the workflow is to calculate albedo using coincident measurements obtained from upward and downward facing pyranometer and quantum measurements, respectively. Upward and downward hemispherical reflectance are measured using broadband pyranometers, the ratio of which provides an estimate of the total shortwave albedo. Similarly, upward and downward hemispherical PAR are measured using quantum sensors and the ratio is an estimate of the visible albedo. Here, downward radiation is measured at a fixed ground station and it is recommended to keep the ground unit within the vicinity to keep the upward and downward facing sensors under the same illumination condition, especially when the illumination conditions rapidly vary with variable cloud cover.

3. Test Measurements at Mer Bleue Wetland

A test case study was conducted in Mer Bleue wetland area (Figure 5) which is a boreal peatland ecosystem that is commonly found in northern Canada.

3.1. Study Area

The Mer Bleue Conservation Area (Figure 5) is a 33.43 km^2 protected area (45°24′34″N, 75°31′07″W) in Eastern Ontario, Canada. The conservation area is about 70 m elevation from mean sea level and is covered mostly by bog, marsh and bordered by patches of forests. The sphagnum bog contains treed bog (black spruce forest) and the open bog vegetation. The species composition of wetland types are described based on [26]. The forest is dominated by black spruce (*Picea mariana*) with some larch (*Larix laricina*), trembling white aspen (*Populus tremuloides*) and grey or white birch (*Betula spp.*). The bog has a complete ground cover of mosses (*Sphagnum capillifo-lium, Sphagnum magellanicum*), with a shrub canopy dominated by evergreen shrubs (*Chamaedaphne calyculata, Kalmia angustifolia,* and *Ledum groenlandicum*), with some deciduous shrubs (*Vaccinium myrtilloides*) and scattered sedges. The marsh areas around Mer Bleue are covered by plants such as cattails (*Typha latifolia*), alders (*Alnus rugosa*), willows (*Salix spp.*), and a variety of sedges (*Carex spp.*). The upland area at the boundary of the bog is covered with mixed forest of conifers and deciduous species. The predominant species in the mixed forest included white pine (*Pinus strobus*), hemlock (*Tsuga canadensis*), sugar maple (*Acer saccharum*) and beech (*Fagus spp.*).

3.2. UAV Data Acquisition

Flight planning was done with the Litchi for DJI (app), which is used to execute full autonomous flight. The UAV is then controlled automatically with the pre-defined mission path according to the given flying altitude of 30 m and waypoint coordinates. Five flights (3 flights on 30 August 2017 and 2 flights on 28 September 2017) with different UAV mission paths have been tested (Figure 5). All five flights were operated between 1:00pm to 3:00pm under variable cloud conditions. Out of 5, four were planned to cover small area with low speed (7 km/h) suitable for mosaic, point cloud and sensor measurements purposes. The overlap is more important for the creation of point clouds than the mosaics and along path overlap exceeded 90% while the across overlap was set to more than 60% for accurate photogrammetric processing. A mission path (Flight 5) was designed to collect sensor measurement and mosaic over large area (about a MODIS pixel size ~500 m) within the battery life (about 17 min). In this survey, the relatively high speed (20 km/h), sometimes resulted in UAV's roll and pitch exceeding the limit (roll and pitch <3°). The cosine-response error increases as solar zenith angle increases [14] so measurements were acquired as close as possible to solar noon. Flying altitude is optimized to 30 m above ground level to flyover the forest as well as to sample a large footprint. Further, a homogenous bog area was monitored from different heights under clear sky or cloudy conditions. All the acquired data were processed following the standard processing chain described in Section 2.3 and the land surface parameters such as total shortwave albedo from pyranometer and visible albedo from quantum sensor were estimated.

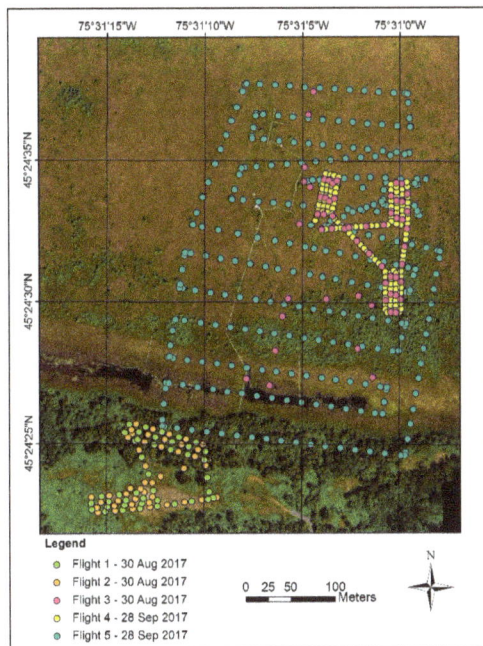

Figure 5. Mer Bleue wetland in eastern Ontario near Ottawa (45°24′34″N, 75°31′07″W). Measurement locations of 5 flight paths are in different colors and highlight difference in line spacing. Cyan path (Flight 5) covers an area of approximately a 500 m MODIS pixel.

3.3. Spectrometer Albedo Estimation

Total shortwave albedo was estimated from spectrometer surface reflection measurements. Usually, spectrometer albedo estimation underestimates surface albedo [27] because the spectrometer cannot record the anisotropic characteristics of land surface. Since the pyranometer measures the hemispherical reflectance and contains the information of surface anisotropy [14], albedo from pyranometer measurements can be used to correct surface reflectance measured by the spectrometer. The calculation of albedo from spectrometer measurements was done in four steps: (1) the ratio of pyranometer albedo to the spectrometer reflectance (for both Green and NIR bands) within the footprint of pyranometer was calculated; (2) the mean ratios of all pyranometer sampling points on two dates, August 30 and September 28, were calculated; (3) the reflectance (Green/NIR) for each sampling point was multiplied by the mean ratio to get the spectral albedo in Green and NIR bands; (4) the empirical equation (Equation (2)) by [28] was used to convert the spectral albedo in Green and NIR bands to broadband shortwave albedo.

$$\alpha = 0.726 \cdot \alpha_{Green} - 0.322 \cdot \alpha_{Green}^2 - 0.051 \cdot \alpha_{NIR} + 0.581 \cdot \alpha_{NIR}^2 \tag{2}$$

3.4. Satellite Based Albedo Estimation

Satellite-based albedo was estimated using Landsat 8 and Sentinel-2 data for comparison with the UAV-based albedo measurements. Ten Landsat 8 [29] OLI Level 2 (L2) images throughout the growing seasons were downloaded from earthexplorer.usgs.gov and met the requirements of being predominately cloud-free. The Landsat 8 OLI data is of high radiometric quality and 30 m spatial resolution. Landsat 8 L2 data are atmospherically corrected 'Science Products', using the Landsat Surface Reflectance Code (LaSRC) to give surface reflectance. This atmospheric correction algorithm differs from previous Landsat 4/5 TM data which used the 6S radiative transfer model. Isolated cumulus clouds are visible but the images could still be used in conjunction with the UAV flights, where the relevant pixels are cloud-free.

Sentinel-2 [30] is the newest generation Earth Observation (EO) satellite which has 13 spectral bands with 10 to 60 m spatial resolution. This is an advantage of higher spatial and spectral resolutions over its counterpart Landsat 8 OLI. Six Sentinel-2 Level-1C (L1C) Top of Atmosphere (TOA) images that are closest to the time of UAV flights and with cloud cover less than 10% were downloaded from the European Space Agency (ESA) Copernicus Open Access Hub (https://scihub.copernicus.eu/dhus/#/home). The Sen2Cor processor was used to perform atmospheric, terrain, and cirrus correction of the L1C product to produce the ortho-image Bottom-Of-Atmosphere (BOA) reflectance product (Level-2A). Since not all the bands are at the same spatial resolution, all bands were atmospherically corrected at 60 m and resampled to 30 m in order to match the spatial resolution of the Landsat images. SNAP software was used to convert the product to a GeoTIFF.

BRDF-based algorithms for estimating broadband surface albedo/total shortwave albedo from satellite observations involve an explicit procedure for the spectral and angular integration of reflectance data. However, due to the narrow field of view of Landsat 8 and Sentinel-2 sensors, it is not possible to acquire directional reflectance under different solar-view geometry and therefore they fail to capture the anisotropy characteristics of most land surfaces [31]. Many BRDF-based methods use external BRDF information from the 8-day 500 m MODIS BRDF model parameter product (MCD43A1) [10] to compute the directional reflectance across all viewing and solar zenith angles, for a given scene [32]. The integrated shortwave MODIS BRDF isotropic, volumetric and geometric parameters were used to calculate surface reflectance at the view and solar zenith angles of each Landsat 8 and Sentinel 2 scene [33] (Equation (3)):

$$A = (\alpha/r(\Omega_l)) \cdot r_l \tag{3}$$

where A is Landsat/Sentinel narrowband albedo to be calculated, r_l is observed Landsat/Sentinel reflectance, Ω_l is viewing and solar geometry of Landsat/Sentinel data, α is albedo, and $r(\Omega_l)$ is the reflectance at Landsat/Sentinel sun view geometry. Both α and $r(\Omega_l)$ are derived by the MODIS BRDF parameters [33].

The narrowband surface albedo calculated using the algorithm above is then used to calculate total shortwave albedo using the coefficients (Equation (4)) formulated by [34]. The weights were originally developed for Landsat 5 and 7, and converted to Sentinel-2 bands (Equation (5)) by [35], where b_n represents the sensor band number.

$$\alpha Liang_{L8} = 0.356 \cdot \rho_{b2} + 0.130 \cdot \rho_{b4} + 0.373 \cdot \rho_{b5} + 0.085 \cdot \rho_{b6} + 0.072 \cdot \rho_{b7} - 0.0018 \tag{4}$$

$$\alpha Liang_{S2} = 0.356 \cdot \rho_{b2} + 0.130 \cdot \rho_{b4} + 0.373 \cdot \rho_{b8A} + 0.085 \cdot \rho_{b11} + 0.072 \cdot \rho_{b12} - 0.0018 \tag{5}$$

Whilst the weighting functions in Equations (4) and (5) were developed for Landsat 5 TM bands, corresponding bands over the same spectral range from Landsat 8 and Sentinel-2 data have similar characteristics, including centre wavelength position and the full width half maximum. Nonetheless, this may impact on the overall absolute accuracy of derived albedo products and their agreement [35].

3.5. Scaling between Observations

The field of view (FOV) specifications vary considerably between the UAV mounted camera and sensors and the satellite sensors, which leads to differences in the measured instantaneous field of view or ground sampling footprint. Table 1 gives an overview of the spatial resolution sampled by the instruments.

Table 1. Field of view (FOV) and ground sampling specifications of the UAV and Satellite sensors.

UAV/Satellite Sensors	Flight Altitude	FOV	Ground Footprint (Diameter)
UAV camera	30 m	-	1.84 cm
UAV spectrometer	30 m	25°	13.3 m
UAV pyranometer/ UAV quantum sensor	30 m	180° – true FOV 172° – restricted FOV 90° – restricted FOV	Infinite 858 m 60 m
Sentinel-2	-	-	20 m
Landsat 8 OLI	-	-	30 m

The comparative differences between the sampling footprints for the UAV mounted instruments versus Landsat 8 OLI 30 m pixels are shown in Figure 6. As the pyranometer FOV of 180°, which gives a sampling footprint reaching infinity, is not practical for comparison against other albedo products, we use a FOV of 90°. This FOV represents 50% of the contributing ground albedo signal [36]. The spectrometer has a FOV of 25°, which leads to a ground sampling footprint radius of 13.3 m and is comparable to high-resolution pixel size (Figure 6).

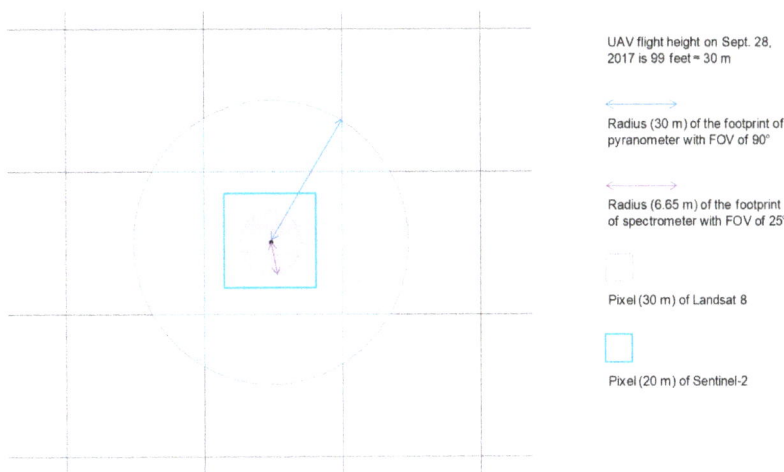

UAV flight height on Sept. 28, 2017 is 99 feet ≈ 30 m

Radius (30 m) of the footprint of pyranometer with FOV of 90°

Radius (6.65 m) of the footprint of spectrometer with FOV of 25°

Pixel (30 m) of Landsat 8

Pixel (20 m) of Sentinel-2

Figure 6. The field of view (FOV) specifications and on-ground sampling footprint of the UAV-mounted spectrometer (6.65 m radius) and pyranometer (30 m radius) in comparison to 30 m Landsat 8 and 20 m Sentinel-2 pixels.

4. Results

4.1. Orthomosaic and Point Cloud

An orthomosaic and point cloud for the Mer Bleue wetland from a mission on 28 September 2017 provides a characterization of the area (Figure 7). The orthomosaic with 1cm spatial resolution provides information on surface texture and land cover type with high accuracy at locations where surface albedo was measured. The point cloud contains 24.5 million points over 1.85 ha and are visualized with natural (visible) colors and color coded for height. All points in the cloud are between 69.8 m (blue) and 75.4 m (red) above sea level (ASL). This gives detailed vegetation density and structural information. Vegetation at the Earth's surface has various levels of structural formation. Open canopies (grass, marsh, open bog) generally have simple structures with leaves more or less randomly distributed in space, whereas foliage in closed canopies (treed bog, forest) are often organized in structures at various hierarchical levels, such as shoots, branches, whorls, tree crowns, and tree groups [37]. This structural information is useful to understand and interpret surface anisotropic and BRDF characteristics, which directly influence surface albedo estimates.

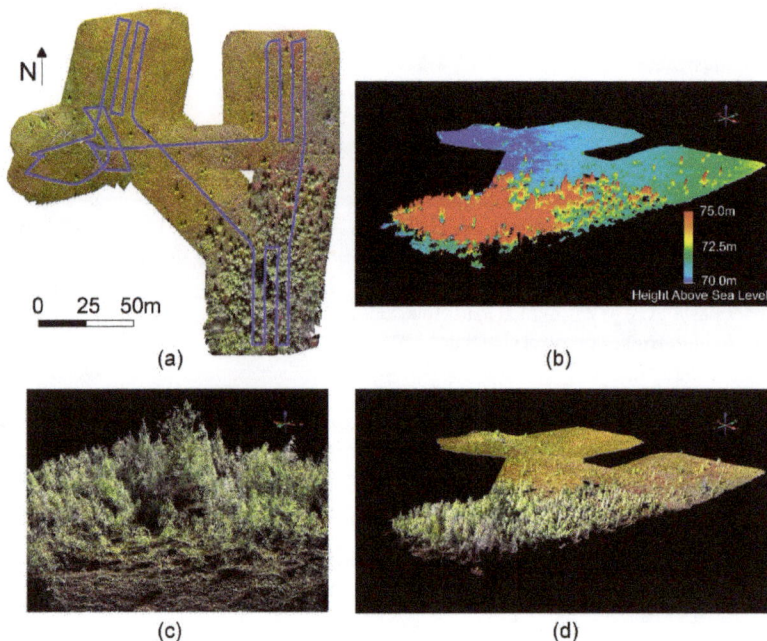

Figure 7. (a) Orthomosaic, (b) height above sea level and (c,d) point cloud showing land cover and vegetation structure of Mer Bleue Bog area (28 September 2017—Flight 4).

4.2. Total Shortwave Albedo from Pyranometer

The total shortwave albedo measurements have been made using instruments such as pyrometers, positioned either on tripods (~3 m in height) or on towers, of 20 m or greater in height [38]. With this stable UAV system, we could measure the surface albedo at different heights (~3 m and higher). The surface albedo measured using the UAV pyranometer is not influenced by the observation height where the land surface is homogenous. Total shortwave albedo measurements over a homogenous bog area at different heights under clear sky or cloudy conditions gives very consistent albedo values (Figure 8a). Further, total shortwave albedos under clear-sky conditions and those under cloudy conditions are almost equal and stable (see Figure 8a,b). Liang et al. [39] stated that under cloudy conditions, the total shortwave albedo is different from that of the clear-sky conditions since the spectral distributions of the downward irradiance at the surface are different. In this study, the differences (Figure 8a,b) were found to be insignificant, which is partially due to minimal atmospheric influences in the measurements.

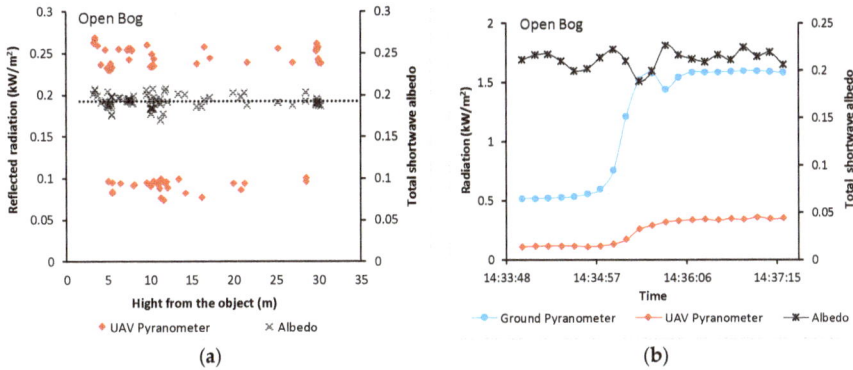

(a) (b)

Figure 8. Open bog total shortwave albedo (**a**) with changing UAV pyranometer height over homogenous open bog under clear sky and cloudy conditions (**b**) from Flight 3 (30 m height) observations over open bog while changing from cloudy to clear sky conditions.

Thus, the land cover appears to dominate the spatial variability of total shortwave albedo in the Mer Bleue wetland, having a highest mean albedo for grass (0.2) and lowest mean albedo for marsh (0.15) (Figure 9). The standard deviation of the five land cover classes have a standard deviation increase with declining mean value (Figure 9). The treed bog has a lower mean albedo with higher standard deviation compare to open bog and forest because isolated trees increase shading effect (see Figure 7). Additionally, in some places marsh appear to be darker (see Figure 5) because of the water level and open water gaps that brings the albedo value lowest with high standard deviation.

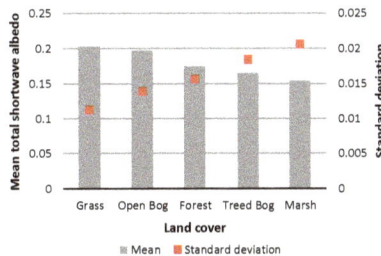

Figure 9. Mean and standard deviation of total shortwave albedo of different land cover types.

4.3. Visible Albedo from Quantum Sensor

The differences in total shortwave albedo and visible albedo from different land covers can be resolved with some overlap present (Figure 10). Land cover with high green leaf coverage has relatively high total shortwave albedo because of high NIR reflection and low visible albedo as the visible energy is absorbed for photosynthesis. Marsh with dry cattails exhibits relatively high total shortwave albedo compared to wet marsh due to the NIR reflectance as well as high visible albedo due to their low photosynthetic activity. Some of the treed bog have low total shortwave albedo and visible albedo values as a result of the shadow casted by isolated trees. However, based on the measurements from all the landcovers of Mer Bleue wetland, visible albedo from the quantum sensor is about one-fourth of the total shortwave albedo from the pyranometer, because of the difference in spectral response functions (see Figure 2).

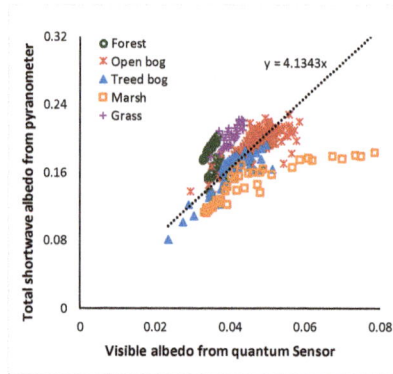

Figure 10. Visible albedo from quantum sensor vs total shortwave albedo from pyranometer.

4.4. Hyperspectral Reflectance and Total Shortwave Albedo from Spectrometer

Measurements of vegetated land surfaces are a mixture of different components including dominant plants, other plants, soils, water, shadows, etc [40]. Based on the surface heterogeneity the land surfaces absorb and reflect different amounts of energy at different wavelengths that can be characterized by their spectral response pattern. Due to the growing demand on more accurate prediction of land surface properties for advanced models based on remotely sensed data, spectral response pattern of different land cover types are very demanding. From the hyperspectral reflectance measurements, one can calculate any user-defined vegetation or surface indices [41]. Indices have been used widely for various land surface characterization, such as the Normalized Difference Vegetation Index (NDVI) as an indicator of green biomass, the Photochemical Reflectance Index (PRI) as a good predictor for photosynthetic efficiency or related variables [42], among others. Moreover, hyperspectral reflectance can be converted to total shortwave albedo as described in the methodology. A comparison between total shortwave albedo directly measured from pyranometer and that derived from spectrometer highlights the separation and overlap in response (Figure 11).

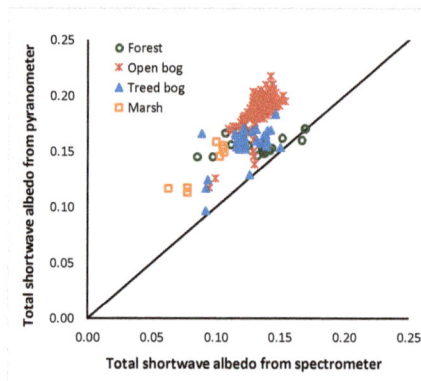

Figure 11. Comparison between total shortwave albedo directly measured using pyranometer and that derived from spectrometer.

The difference between the total shortwave albedo from the pyranometer and spectrometer may be because of 3 reasons: (1) spectral response function of LI-200R Pyrometer, (2) the empirical equation (Equation (2)), which was used to derive the total shortwave albedo of vegetation from spectrometer data, was based on a study of glaciers, and (3) difference in their FOVs. This bias is observable in the data where the pyranometer data has higher values than the spectrometer data (Figure 11). The pyranometer and spectrometer instruments are mounted on board the same platform; however, the differences in their FOVs introduce variations in the ground area that is measured by each sensor. The spectrometer-derived albedo estimates are aggregated to 60 m footprint diameter, which represents about 50% of the actual ground albedo contribution that is measured by the pyranometer. Due to limited spatial extent sampled during the UAV flight, it was not feasible to 'upscale' the spectrometer albedo measurements to represent 80% of the pyranometer FOV (850 m) because it would have included all spectrometer albedo ground points.

4.5. Comparison UAV-Derived Albedo with Satellite-Derived Albedo

The surface albedo estimates from Landsat 8 and Sentinel-2 satellite data with BRDF correction have good agreement over the Mer Bleue wetland for dates closes to a UAV flight (Figure 12). Our analysis has indicated that the Landsat 8 and Sentinel-2 albedo products demonstrate a strong linear relationship (R^2 = 0.82, not shown).

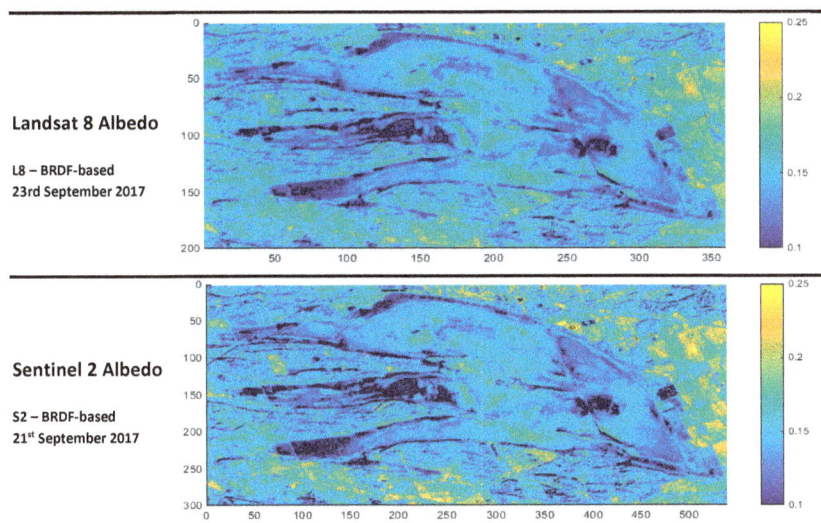

Figure 12. Spatial total shortwave albedo images (pixel coordinates) from Landsat 8 and Sentinel-2 satellites.

Figure 13 shows a comparison between the total shortwave albedo derived from Landsat 8 and Sentinel-2 and the total shortwave albedo from the pyranometer measurements during the flight on 28 September 2017. The Landsat 8 and Sentinel-2 albedo estimates were aggregated to match the UAV-pyranometer footprint (90° FOV, 60 m diameter) for direct comparison between the two estimates. Theoretically, FOV of the pyranometer is 180° and thus the footprint is infinite. However, the cosine correction is performed for up to 82° of the incidence angle for Li-200R, and 50% of the signal comes from a FOV of 90° and 80% from a FOV of 127° [36]. Therefore, as the average height of the UAV flight is 30 m, 50% of the signal to the downward-facing pyranometer comes from a footprint diameter of 60 m, which is used for the comparison. Data points acquired during cloudy periods were excluded from the analysis, because, variable cloud presence may lead to shadows cast on the ground-based pyranometer, which may lead to low measurements of downward irradiance. This results in erroneously increase albedo calculated from the ratio of the upwelling to the downwelling irradiance.

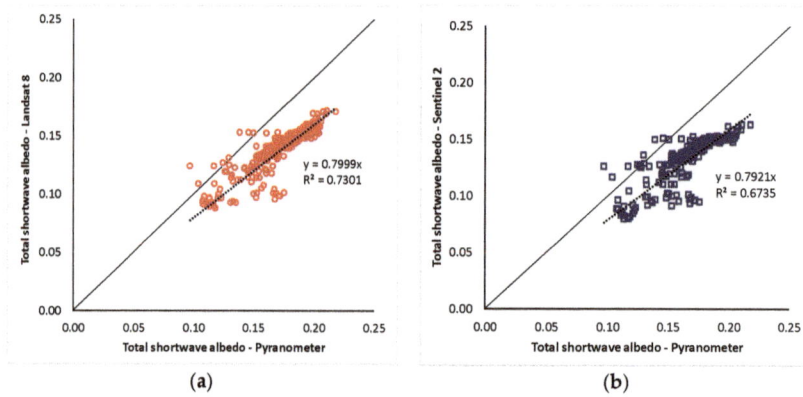

(a) (b)

Figure 13. Comparison between satellite-derived albedo using the BRDF-based approach for (a) Landsat 8, and (b) Sentinel-2 sensors, against directly measured albedo from pyranometer.

There is a relatively strong relationships between satellite-derived albedo and directly measured albedo from the pyranometer, across all land cover types (Figure 13; R^2 = 0.73 for Landsat 8 and R^2 = 0.68 for Sentinel-2). There is a linear bias with satellite-derived albedo underestimated by 20% relative to the values from the pyranometer. The spectral response function of the pyranometer, bias in the satellite-derived albedo estimation and FOVs could explain this (see Sections 5.3–5.5).

The satellite-derived albedo is compared to spectrometer-derived albedo (Figure 14). The spectrometer-derived albedo is derived at ground sampling footprints of 13.3 m in diameter, and the Landsat 8 and Sentinel-2 albedo estimates are for the original satellite resolutions of 30 m and 20 m, respectively. The results are for the UAV data sampled during the September 28th flight. The satellite-derived albedo estimates and the spectrometer-derived albedo scattered along the 1:1 line. Variations in the relationship between the satellite-derived and spectrometer-derived albedo estimates may also arise from differences in the band characteristics and the spectral range that reflectance data are collected over. Most notably, the spectrometer lacks two SWIR bands that the two satellite sensors possess, which led to a different method used for spectrometer based albedo estimates.

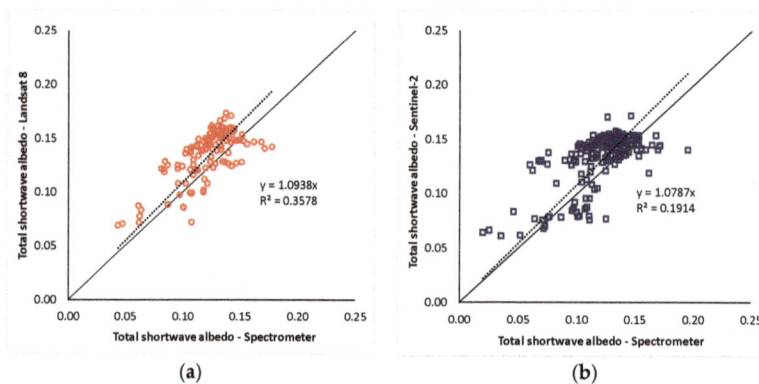

(a) (b)

Figure 14. Comparison between satellite-derived total shortwave albedo using the BRDF-based approach for (a) Landsat 8, and (b) Sentinel-2 sensors, against spectrometer-derived total shortwave albedo.

5. Discussion

5.1. Operational UAV Concerns

The UAV sensor system has the potential to characterize a land surface, however, field data acquisition with the UAV sensor system faces several minor challenges such as: limited battery life; slow flight to maintain flight stability; calibration requirements; synchronizing UAV and all the sensors of the UAV and the ground system; and acquiring required permits. Further, the weather conditions such as wind and temperature need to be suitable for stable flight and sensor measurements. Variable clouds could affect the consistencies in the radiation environment between the actual footprint of the UAV sensors and the ground station sensors.

5.2. Issues Related to Mini-Sensors

Compared to standard commercial pyranometers and hyperspectral sensors, small sensors are lightweight, generally more easily mounted on UAV, easily replaced, and cost effective, which allows us to use multiple sensors. We expected some variation between sensors. For example, Li-Cor sensors do not have spectral response function for each individual sensor. However, we performed an inter-comparison between the identical sensors by measuring variable downwelling irradiance at the same time and the result had very good 1:1 correlation. Another issue is that the spectral reflectance of VNIR is from 2 sensors, VIS and NIR, middle portions of the spectrum avoiding the edges from two sensors were merged together. SWIR measurements are useful for various application mainly for estimating total shortwave albedo. But, Ocean optics does not produce small SWIR sensor suitable for UAV.

5.3. Spectral Response Functions

The Li-Cor pyranometer is small in size and mass, however, the limited spectral sensitivity (400 nm to 1100 nm) of the pyranometer is a disadvantage [43]. In the case of snow, the Li-Cor 200SZ reflected radiation data with a positive bias when compared to standard optical black thermopile instruments that have a nearly uniform response in the full ultraviolet to infrared (285–2800 nm) range [43]. Whilst the Li-Cor pyranometer samples broadband reflectance, rather than in discrete narrow bands, its spectral response function has a higher contribution of sampled irradiance from NIR (Figure 2). However, even more importantly, the pyranometer only measured visible-NIR wavelengths and does not cover SWIR region. This may have a disproportionately high effect if there are variations in vegetation or soil water content between land covers, or over space and time. In only sampling to the NIR, this implies an inherent assumption that there will be little variation across the remainder of the solar spectrum. This uncertainty in the total surface albedo measurements could be one of the main sources of linear bias in the results, with satellite-derived albedo underestimated, relative to the pyranometer results.

5.4. Satellite Based Albedo Estimation

There are a number of well-used empirical coefficients that have been used to convert satellite measured spectral reflectance or radiance into total shortwave albedo [34,44]. However, most of the empirical equations (including Equations (4) and (5)) are based on wide range of albedo data from various landcovers with different shape of spectral reflectance, compared to the vegetation cover in the study are. Further, these empirical methods sometimes neglect to take into consideration the surface anisotropy of the land cover present. Naegeli et al. [35] found that surface anisotropy accounts for an underestimation of albedo of up to 10%, depending on the surface type. Darker surfaces are more sensitive to anisotropy correction compared to brighter surfaces [35]. Furthermore, the time of data acquisition may also lead to uncertainties in the retrieved albedo product [14].

5.5. Field of View

In previous studies, validation of satellite-retrieved albedo using directly measured albedo from radiation sensors was usually conducted with fixed, ground or tower-mounted instruments [38]. The novel use of an on board, UAV-based sensors allows direct albedo measurements in a spatially-continuous manner and provides opportunity to validate satellite estimates using more validation points and, importantly a variety of land covers. Other than that, by averaging multiple UAV-derived albedo values within one pixel of remote sensing image, the issue of upscaling from point measurement to the pixel level can be alleviated. In this case, 50% of contributions to the pyranometer albedo are coming from outside of the 60 m diameter footprint that is compared with aggregated Landsat 8 and Sentinel-2 60 m pixels. Greater uncertainty and error will arise when this is within a heterogeneous landscape, and particularly in landscapes where the albedos from different land covers are very different from each other, such as treed bog adjacent to open bog.

6. Conclusions

This research evaluated the use of a UAV-system for deriving and monitoring surface albedo over a wetland ecosystem for operational hydrological monitoring at fine spatial-scales and short temporal intervals. Pairs of low-cost and lightweight VIS and NIR spectrometers, pyranometers and quantum sensors were used on the ground and on the UAV for measuring downwelling irradiance and upwelling radiance, respectively. Videos from UAV mounted camera was used for extracting land surface structural information and reflectance from spectrometers were used for categorising land surface spectral characteristics. Comparison of total shortwave albedo from spectrometers and total shortwave albedo from pyranometer, allowed quantitative evaluation of the direct albedo from pyranometer against Landsat 8 and Sentinel-2 derived albedo. There were relatively little differences in total shortwave albedo derived from Landsat 8 and Sentinel-2 images, likely due to their similar spectral response functions and spatial resolution, with both satellite results having a strong performance with direct total shortwave albedo measurements from UAV pyranometer (R^2 = 0.73 and R^2 = 0.68, respectively). However, there is a 20% underestimation that could be due to limited spectral range of LI-200R, generalized use of empirical satellite albedo retrieval equations, and/or the restricted 50% pyranometer FOV that is used for comparison with the satellite pixels. This study demonstrates the potential of a UAV platform for bridging the gap between fixed point, in-situ albedo measurements and pixel-level measurements by satellites, for producing accurate, time sensitive and fine-scale spatially continuous albedo measurements for hydrological monitoring over sensitive and dynamic wetland environments.

Author Contributions: Conceptualization, F.C. and S.W.; Data curation, F.C., S.G.L. and R.W.; Formal analysis, F.C., S.G.L., H.C. and R.W.; Funding acquisition, S.W. and H.A.J.R.; Investigation, F.C., S.W., H.C., S.G.L. and J.C.; Methodology, F.C., S.W., H.C., J.C. and R.W.; Project administration, S.W. and H.A.J.R.; Writing—Original Draft, F.C.; Writing—Review & Editing, S.W., H.C., H.A.J.R. and R.W.

Funding: The UAV-Sensor system has been developed with financial-support from Natural Resources Canada Canada Centre for Mapping and Earth Observation - Water Program; and Geological Survey of Canada - Ground Water Geoscience, and Climate Change Geoscience programs.

Acknowledgments: The authors gratefully acknowledge the support from the CCRS management team for this activity. We are also immensely grateful to Alexander Trichtchenko and Fuqun Zhou for their comments on an earlier version of the manuscript. We thank the editors and anonymous reviewers, who provided scientific and editorial peer review of the manuscript and helpful comments to improve the content.

Conflicts of Interest: The authors declare no conflict of interest.

References

1. Wang, S.; Trishchenko, A.P.; Sun, X. Simulation of canopy radiation transfer and surface albedo in the EALCO model. *Clim. Dyn.* **2007**, *29*, 615–632. [CrossRef]
2. Franch, B.; Vermote, E.; Claverie, M. Intercomparison of Landsat albedo retrieval techniques and evaluation against in situ measurements across the US SURFRAD network. *Remote Sens. Environ.* **2014**, *152*, 627–637. [CrossRef]
3. Carter, G.A. Responses of leaf spectral reflectance to plant stress. *Am. J. Bot.* **1993**, *80*, 239–243. [CrossRef]
4. Davidson, A.; Wang, S. The effects of sampling resolution on the albedos of dominant land cover types in the North American boreal region. *Remote Sens. Environ.* **2004**, *93*, 211–224. [CrossRef]
5. Davidson, A.; Wang, S. Spatio-temporal variations in land surface albedo across Canada from MODIS observations. *Can. J. Remote Sens.* **2005**, *31*, 377–390. [CrossRef]
6. Hanesiak, J.M.; Stewart, R.E.; Bonsal, B.R. Characterization and summary of the 1999-2005 Canadian prairie drought. *Atmos. Ocean* **2011**, *49*, 1–32. [CrossRef]
7. Wang, S.; Trishchenko, A.P.; Khlopenkov, K.V.; Davidson, A. Comparison of International Panel on Climate Change Fourth Assessment Report climate model simulations of surface albedo with satellite products over northern latitudes. *J. Geophys. Res.* **2006**, *111*, D21108. [CrossRef]
8. Betts, A.K.; Ball, J.H. Albedo over the boreal forest. *J. Geophys. Res.* **1997**, *102*, 28901–28910. [CrossRef]
9. Qu, Y.; Liang, S.; Liu, Q.; He, T.; Liu, S.; Li, X. Mapping surface broadband albedo from satellite observations: A review of literatures on algorithms and products. *Remote Sens.* **2015**, *7*, 990–1020. [CrossRef]
10. Schaaf, C.B.; Gao, F.; Strahler, A.H.; Lucht, W.; Li, X.; Tsang, T.; Strugnell, N.C.; Zhang, X.; Jin, Y.; Muller, J.P.; et al. First operational BRDF, albedo nadir reflectance products from MODIS. *Remote Sens. Environ.* **2002**, *83*, 135–148. [CrossRef]
11. Trishchenko, A.P.; Luo, Y.; Khlopenkov, K.V.; Wang, S. A method to derive the multi-spectral surface albedo consistent with MODIS from historical AVHRR and VGT satellite data. *J. Appl. Meteorol. Climatol.* **2008**, *47*, 1199–1221. [CrossRef]
12. Maignan, F.; Bréon, F.M.; Lacaze, R. Bidirectional reflectance of Earth targets: Evaluation of analytical models using a large set of spaceborne measurements with emphasis on the Hot Spot. *Remote Sens. Environ.* **2004**, *90*, 210–220. [CrossRef]
13. Geiger, B.; Carrer, D.; Franchistéguy, L.; Roujean, J.L.; Meurey, C. Land surface albedo derived on a daily basis from Meteosat Second Generation observations. *IEEE Trans. Geosci. Remote Sens.* **2008**, *46*, 3841–3856. [CrossRef]
14. Ryan, J.C.; Hubbard, A.; Box, J.E.; Brough, S.; Cameron, K.; Cook, J.M.; Cooper, M.; Doyle, S.H.; Edwards, A.; Holt, T. Derivation of high spatial resolution albedo from UAV digital imagery: Application over the Greenland Ice Sheet. *Front. Earth Sci.* **2017**, *5*, 40. [CrossRef]
15. Adão, T.; Hruška, J.; Pádua, L.; Bessa, J.; Peres, E.; Morais, R.; Sousa, J.J. Hyperspectral imaging: A review on UAV-Based Sensors, Data Processing and Applications for Agriculture and Forestry. *Remote Sens.* **2017**, *9*, 1110. [CrossRef]
16. Burkhart, J.F.; Kylling, A.; Schaaf, C.B.; Wang, Z.; Bogren, W.; Storvold, R.; Solbø, S.; Pedersen, C.A.; Gerland, S. Unmanned aerial system nadir reflectance and MODIS nadir BRDF-adjusted surface reflectances intercompared over Greenland. *Cryosphere* **2017**, *11*, 1575. [CrossRef]
17. Rippin, D.M.; Pomfret, A.; King, N. High resolution mapping of supra-glacial drainage pathways reveals link between micro-channel drainage density, surface roughness and surface reflectance. *Earth Surf. Process. Landf.* **2015**, *40*, 1279–1290. [CrossRef]
18. Ryan, J.C.; Hubbard, A.; Stibal, M.; Box, J.E. Attribution of Greenland's ablating ice surfaces on ice sheet albedo using unmanned aerial systems. *Cryosphere Discuss.* **2016**. [CrossRef]
19. DJI. Matrice 600 User Manual DJI. 2016. Available online: https://dl.djicdn.com/downloads/m600/20170717/Matrice_600_User_Manual_v1.0_EN.pdf (accessed on 1 February 2017).
20. Li-Cor. Principles of Radiation Measurement. LI-COR Ltd., 2015. Available online: https://licor.app.boxenterprise.net/s/liuswfuvtqn7e9loxaut (accessed on 5 February 2017).
21. Mac Arthur, A. *Field Guide for the ASD Fieldspec Pro Radiance/Irradiance Measurements in Raw DN Mode*; ASD Raw DN Mode Field User Note; NERC FSF: Edinburgh, UK, 2007.

22. Ocean Optics. *STS Developer's Kit—Installation and Operation Manual*; Ocean Optics Inc.: Largo, FL, USA, 2014; Available online: http://oceanoptics.com/wp-content/uploads/STSDevKitIO.pdf (accessed on 27 March 2017).

23. OceanView. *OceanView Installation and Operation Manual*; Ocean Optics Inc.: Largo, FL, USA, 2013; Available online: http://oceanoptics.com/wp-content/uploads/OceanViewIO.pdf (accessed on 24 March 2017).

24. Westoby, M.J.; Brasington, J.; Glasser, N.F.; Hambrey, M.J.; Reynolds, J.M. Structure-from-Motion photogrammetry: A low-cost, effective tool for geoscience applications. *Geomorphology* **2012**, *179*, 300–314. [CrossRef]

25. Zeng, C.; King, D.J.; Richardson, M.; Shan, B. Fusion of Multispectral Imagery and Spectrometer Data in UAV Remote Sensing. *Remote Sens.* **2017**, *9*, 696. [CrossRef]

26. Frolking, S.; Roulet, N.T.; Moore, T.R.; Lafleur, P.M.; Bubier, J.L.; Crill, P.M. Modeling seasonal to annual carbon balance of Mer Bleue Bog, Ontario, Canada. *Glob. Biogeochem. Cycles* **2002**, *16*, 4-1–4-21. [CrossRef]

27. Knap, W.H.; Reijmer, C.H. Anisotropy of the Reflected Radiation Field Over Melting Glacier Ice: Measurements in Landsat TM Bands 2 and 4. *Remote Sens. Environ.* **1998**, *65*, 93–104. [CrossRef]

28. Knap, W.; Reijmer, C.; Oerlemans, J. Narrowband to broadband conversion of Landsat TM glacier albedos. *Int. J. Remote Sens.* **1999**, *20*, 2091–2110. [CrossRef]

29. Loveland, T.R.; Irons, J.R. Landsat 8: The plans, the reality, and the legacy. *Remote Sens. Environ.* **2016**, *185*, 1–6. [CrossRef]

30. Drusch, M.; Del Bello, U.; Carlier, S.; Colin, O.; Fernandez, V.; Gascon, F.; Hoersch, B.; Isola, C.; Laberinti, P.; Martimort, P.; et al. Sentinel-2: ESA's Optical High-Resolution Mission for GMES Operational Services. *Remote Sens. Environ.* **2012**, *120*, 25–36. [CrossRef]

31. Roy, D.P.; Wulder, M.A.; Loveland, T.R.; Woodcock, C.E.; Allen, R.G.; Anderson, M.C.; Helder, D.; Irons, J.R.; Johnson, D.M.; Kennedy, R.; et al. Landsat-8: Science and product vision for terrestrial global change research. *Remote Sens. Environ.* **2014**, *145*, 154–172. [CrossRef]

32. Roy, D.P.; Zhang, H.; Ju, J.; Gomez-Dans, J.L.; Lewis, P.E.; Schaaf, C.; Sun, Q.; Li, J.; Huang, H.; Kovalskyy, V. A general method to normalize Landsat reflectance data to nadir BRDF adjusted reflectance. *Remote Sens. Environ.* **2016**, *176*, 255–271. [CrossRef]

33. Wang, Z.; Schaaf, C.B.; Sun, Q.; Kim, J.; Erb, A.M.; Gao, F.; Román, M.O.; Yang, Y.; Petroy, S.; Taylor, J.R.; et al. Monitoring land surface albedo and vegetation dynamics using high spatial and temporal resolution synthetic time series from Landsat and the MODIS BRDF/NBAR/albedo product. *Int. J. Appl. Earth Obs. Geoinform.* **2017**, *59*, 104–117. [CrossRef]

34. Liang, S. Narrowband to broadband conversions of land surface albedo I: Algorithms. *Remote Sens. Environ.* **2001**, *76*, 213–238. [CrossRef]

35. Naegeli, K.; Damm, A.; Huss, M.; Wulf, H.; Schaepman, M.; Hoelzle, M. Cross-Comparison of albedo products for glacier surfaces derived from airborne and satellite (Sentinel-2 and Landsat 8) optical data. *Remote Sens.* **2017**, *9*, 110. [CrossRef]

36. Cescatti, A.; Marcolla, B.; Santhana Vannan, S.K.; Pan, J.Y.; Román, M.O.; Yang, X.; Ciais, p.; Cook, R.B.; Law, B.E.; Matteucci, G.; Migliavacca, M.; et al. Intercomparison of MODIS albedo retrievals and in situ measurements across the global FLUXNET network. *Remote Sens. Environ.* **2012**, *121*, 323–334. [CrossRef]

37. Chen, J.M.; Liu, J.; Leblanc, S.G.; Lacaze, R.; Roujean, J.L. Multi-angular optical remote sensing for assessing vegetation structure and carbon absorption. *Remote Sens. Environ.* **2003**, *84*, 516–525. [CrossRef]

38. Coakley, J.A. Reflectance and albedo, surface. In *Encyclopedia of Atmospheric Sciences*; Holton, J.R., Curry, J.A., Pyle, J.A., Eds.; Academic Press: Cambridge, MA, USA, 2003; pp. 1914–1923.

39. Liang, S.; Strahler, A.H.; Walthall, C. Retrieval of land surface albedo from satellite observations: A simulation study. *J. Appl. Meteorol.* **1999**, *38*, 712–725. [CrossRef]

40. Colwell, J.E. Vegetation canopy reflectance. *Remote Sens. Environ.* **1974**, *3*, 175–183. [CrossRef]

41. Canisius, F.; Fernandes, R. Evaluation of the information content of Medium Resolution Imaging Spectrometer (MERIS) data for regional leaf area index assessment. *Remote Sens. Environ.* **2012**, *119*, 301–314. [CrossRef]

42. Garbulsky, M.F.; Peñuelas, J.; Gamon, J.; Inoue, Y.; Filella, I. The photochemical reflectance index (PRI) and the remote sensing of leaf, canopy and ecosystem radiation use efficiencies. A review and meta-analysis. *Remote Sens. Environ.* **2011**, *115*, 281–297. [CrossRef]

43. Stroeve, J.; Box, J.E.; Gao, F.; Liang, S.L.; Nolin, A.; Schaaf, C. Accuracy assessment of the MODIS 16-day albedo product for snow: Comparisons with Greenland in situ measurements. *Remote Sens. Environ.* **2005**, *94*, 46–60. [CrossRef]

44. Pimentel, R.; Aguilar, C.; Herrero, J.; Pérez-Palazón, M.; Polo, M. Comparison between Snow Albedo Obtained from Landsat TM, ETM+ Imagery and the SPOT VEGETATION Albedo Product in a Mediterranean Mountainous Site. *Hydrology* **2016**, *3*, 10. [CrossRef]

drones

MDPI

Review

Drones for Conservation in Protected Areas: Present and Future

Jesús Jiménez López [1,*] and Margarita Mulero-Pázmány [2,*]

[1] MARE—Marine and Environmental Sciences Centre, Quinta do Lorde Marina, Sítio da Piedade, 9200-044 Caniçal, Madeira Island, Portugal
[2] School of Natural Sciences and Psychology, Liverpool John Moores University, Liverpool L3 3AF, UK
* Correspondence: lopezjimenezjesus@mare-centre.pt (J.J.L.); M.C.MuleroPazmany@ljmu.ac.uk (M.M.-P.)

Received: 30 October 2018; Accepted: 7 January 2019; Published: 9 January 2019

Abstract: Park managers call for cost-effective and innovative solutions to handle a wide variety of environmental problems that threaten biodiversity in protected areas. Recently, drones have been called upon to revolutionize conservation and hold great potential to evolve and raise better-informed decisions to assist management. Despite great expectations, the benefits that drones could bring to foster effectiveness remain fundamentally unexplored. To address this gap, we performed a literature review about the use of drones in conservation. We selected a total of 256 studies, of which 99 were carried out in protected areas. We classified the studies in five distinct areas of applications: "wildlife monitoring and management"; "ecosystem monitoring"; "law enforcement"; "ecotourism"; and "environmental management and disaster response". We also identified specific gaps and challenges that would allow for the expansion of critical research or monitoring. Our results support the evidence that drones hold merits to serve conservation actions and reinforce effective management, but multidisciplinary research must resolve the operational and analytical shortcomings that undermine the prospects for drones integration in protected areas.

Keywords: protected areas; drones; RPAS; conservation; effective management; biodiversity threats

1. Introduction

Protected areas aim to safeguard biodiversity, preserve ecosystem services and ensure the persistence of natural heritage [1]. Despite their essential role in conservation, the allocation of resources to cope with an increasing variety of regular activities and unforeseen circumstances remains generally insufficient [2], severely affecting overall effectiveness [3]. Besides, protected areas subjected to international and national agreements must resolve their acquired responsibilities to maintain their legal status [4]. Hence, there is a demand for cost-effective, versatile and practical initiatives to attend a disparity of requirements to guarantee conservation, including a wide range of natural solutions [5], technological advances, and methods or innovative application of existing technologies [6].

In the last decade, drones (also known as unmanned aerial systems, remotely piloted aircraft systems, RPAS, UAS, UAV) have been the subject of a growing interest in both the civilian and scientific sphere, and indeed avowed as a new distinct era of remote sensing [7] for the study of the environment [8]. Drones offer a relatively risk-free and low-cost manner to rapidly and systematically observe natural phenomena at high spatio-temporal resolution [9]. For these reasons, drones have recently become a major trend in wildlife research [10,11] and management [12–14].

The success of drones can be partially explained by their great flexibility to carry different sensors and devices. The scope of application determines the best combination of aerial platform and payload. Although drones come in many different shapes and sizes, widespread small fixed-wing and rotary-wing aircrafts are frequently used for video and still photography. These consumer grade drones coupled with lightweight cameras and multispectral sensors can deliver professional mapping solutions at a fraction of a cost than previous photogrammetric techniques. Medium size drones can be equipped with compact thermal vision cameras, hyperspectral sensors and laser scanning such as LiDAR, with great prospects for wildlife ecology, vegetation studies and forestry applications respectively [15–17]. Even though visible and multispectral band cameras encompass the most obvious sensing devices, drones can indeed incorporate a diversity of instruments to measure many distinct physical quantities such as temperature, humidity or air pollution [18]. Additionally, large aerial platforms can lift heavier payloads and represent an appropriate solution for integrating complex systems with the capacity to remotely assist sampling, hold cargo or deliver assistance. A brief summary of platforms and sensors is given in Tables 1 and 2 (but see [19–22] for an in-depth revision).

Table 1. Classification of drones according to characteristics and applications.

SIZE									
Nano		Micro		Mini		Small		Medium	Large
<30 mm		30–100 mm		100–300 mm		300–500 mm		500 mm–2 m	>2 m
Maximum Take-Off Weight (MTOW)									
<0.5 Kg		0.5–5 Kg			5–25 Kg			>25 Kg	
RANGE (Distance/Type of Operation)									
Close-range <0.5 miles			Mid-range 0.5–5 miles			Long-range 5 > miles			
Visual Line Of Sight (VLOS)			Extended Visual Line Of Sight (EVLOS)			Beyond Visual Line Of Sight (BVLOS)			
WING									
Rotary wing					Fixed wing				Hybrid (VTOL)
Single Dual rotors	Multi-Rotor				Low Wing	Mid Wing	High Wing	Delta Wing	
	Tricopter	Quadcopter	Hexacopter	Octocopter					
POWER									
Electric			Gas			Nitro		Solar	
ASSEMBLING									
Ready-To-Fly (RTF)			Bind-N-Fly (BNF)			Almost-Ready-to-Fly (ARF)			
APPLICATIONS									
Logistics	Civil Engineering	Disaster Relief	Heritage	Search and Rescue	Precision Agriculture		Natural Resources		Law Enforcement
Wildlife Management	Weather Forecasting	Industrial Inspection	Leisure	Military	Disaster Relief		Aerial Photography and Film		Archeology

Note: SIZE, MTOW and RANGE: based on average values (no specific standard/regulation). ASSEMBLING: level of work required to use the drone since acquisition.

Considering the ample range of possibilities, it is not surprising that some protected areas are adopting drones for various applications. For example, to assist search and rescue [23]; protect endangered turtles from feral species [24]; monitoring invasive plant species [25]; document illegal logging and mining [26]; wetland management [27]; anti-poaching [28]; and marine litter detection [29]. Recently, a team of scientists discovered a biodiversity hotspot using drones [30], which could be argued as a convenient procedure to adequately expand protected areas as established by the Aichi Target 11 [3]. In addition, we are witnessing a continuous development of sophisticated drones and ingenious methods that target particular conservation actions, such as wildfires firefighting [31]; whale health monitoring [32]; disease vectors control [33]; or seed planting for habitat restoration [34]. The fast pace of technological advances and novel applications probably exceeded previous expectations, but also gives rise to singular circumstances that must be placed in the context of management.

Table 2. Summary classification of sensors and devices that can be coupled to drones.

Instrument.		Type of Sensor	Spatial Resolution	Spectral Resolution	Weight	Costs
Imaging sensors	Visible RGB	Passive	Very high 1–5 cm/pixel	Low (3 bands)	Low <0.5 kg	Low $100–1000
	Near Infrared (NIR)	Passive	Very high 1–5 cm/pixel	Low (3 bands)	Low <0.5 kg	Low $100–1000
	Multispectral	Passive	High 5–10 cm/pixel	Medium (5–12 bands)	Medium 0.5–1 kg	Medium $1000–10,000
	Hyperspectral	Passive	High 5–10 cm	High (> 50–100 bands)	Medium 0.5–1 kg	High $10,000–50,000
	Thermal	Passive	Medium 10–50 cm/pixel	Low 1 band	Medium 0.5–1 kg	Medium $1000–10,000
Ranging sensors	Laser scanners (LiDAR)	Active	Very high 1–5 cm/pixel	Low 1–2 bands	High 0.5–5 kg	High $10,000–50,000
	Synthetic Aperture Radars (SAR)	Active	Medium 10–50 cm/pixel	Low 1 band	High >5 kg	Very high >$50,000
Other sensors and devices						
Atmospheric sensors		Temperature, Pressure, Wind, Humidity				
Chemical Sensors		Gas, Geochemical				
Position systems		Ultrasound, Infrared, Radio Frequency, GPS				
Other devices		Recorder device/microphones				
Sampling Devices		Water, Aerobiological, Microbiological Sampling				
Other devices		Cargo, Spraying, Seed spreader				

Some authors have identified negative aspects of drones use in conservation. Potential wildlife disturbance effects [35] need to be further investigated. The use of drones as tools of coercion could weaken the environmental commitment of communities in protected areas [36], and therefore may prove counterproductive for conservation. On the other hand, the massive amount of data acquired with drones require modern, robust and computationally intensive methods to derive accurate and meaningful information [37], which may represent a technological barrier to the effective use of this technology in protected areas.

Likewise, the connection of drone advances with the most important features guiding effective management has not yet been specifically weighted and would be necessary to better align research efforts to conservation priorities. In addition, whether decision makers can take practical advantage of present and oncoming advances in the discipline remains questionable for several reasons. To find early answers to these remarks, we conducted an extensive literature review of drone applications with potential to enhance the effective management of protected areas. This perspective may help identify plausible scenarios where drones can be used in a rational and efficient manner.

2. Methods

We conducted a comprehensive literature search on drones in conservation up to October 2nd 2018, in line with related studies [10,11,35]. All searches were done by the same person in English, mainly using Google Scholar. This was further complemented through reference harvesting, citation tracking, abstracts in conference programs, and author search, using Research Gate and Mendeley (see PRISMA Flowchart in Supplementary Figure S1 Checklist and list of studies reviewed in Table S1). We then removed duplicate and unrelated results. Finally, peer-reviewed publications were collated and revised.

Keywords on the search included drones in their various meanings and acronyms: "unmanned aircraft systems", "UAS", "remotely piloted aerial system", "RPAS", "drone", "model aircraft", "unmanned aerial vehicle", "UAV", "unmanned aircraft system". These were combined with terms referring to threats and common conservation measurements in protected areas: "protected area", "conservation", "ecology", "ecosystem", "habitat", "vegetation", "forest", "wetland", "reforestation", "monitoring", "survey", "sampling", "inventory", "wildlife", "fauna", "bird", "mammal", "fish", "amphibian", "reptile", "wildfire", "landslide", "remote sensing", "tourism", "ecotourism", "law enforcement", "poaching", "anti-poaching", "logging", "risk management", "pollution", and "search and rescue". In total, we applied 47 search terms and combinations using logical disjunctions.

We classified the studies into categories that represent the common threats and essential management measures in protected areas [5,38–40]. The categories are: "wildlife research and management" for those projects aimed at observing wildlife, estimating population parameters such as abundance and distribution, and establishing management measures to mitigate human-wildlife conflicts (n = 96); "ecosystem monitoring" for applications related with the study and mapping of natural habitats (n = 106); "Law enforcement" encompassing poaching and other illicit activities (n = 6); "Ecotourism" referring to recreational activities and visitors management (n = 3); "Environmental management and emergency response" spanning environmental monitoring and protection, natural hazards, search and rescue operations and similar cases (n = 45). We briefly tackled legal and ethical issues, including potential impact on wildlife and habitats, but also economic and technological factors, since all shape the feasibility of drones to approach conservation and environmental issues.

3. Results and Discussion

The literature search on drones in conservation provided a total of 256 studies. Of these, 99 describe applications that were accomplished in terrestrial and marine protected areas, according to the Protected Planet database [41]. The typology of protected areas includes national, international designations and registered private initiatives, with all UICN management categories (Ia, Ib, II, III, IV, V, VI) represented [1]. We found examples on all continents and in most ecosystems. The United States of America lead the ranking of countries where more drone studies have taken place (45), followed by Canada (26), Australia (17), China (11), Germany (11) and Spain (9). Figure 1 summarizes the selected research.

A   Location of collected Drone Studies grouped at Country Level and Drone Studies in Protected Areas

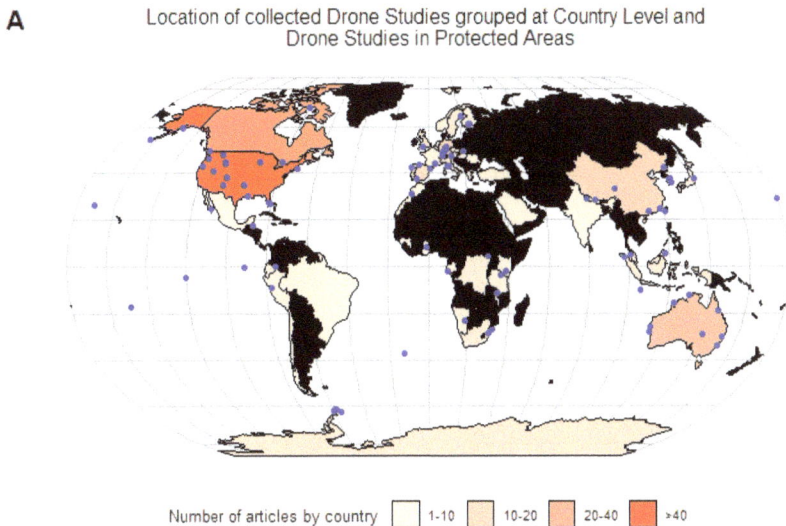

Number of articles by country ☐ 1-10 ☐ 10-20 ☐ 20-40 ☐ >40

Figure 1. *Cont.*

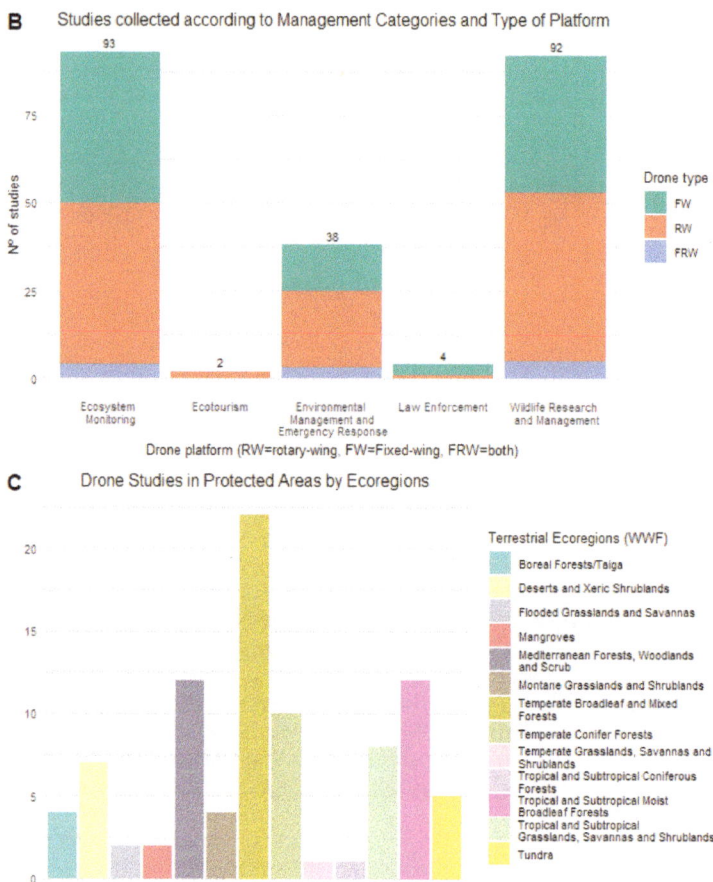

Figure 1. (**A**) Blue points represent studies in protected areas. Choropleth map shows location of studies by country. No studies were collected in countries colored black. (**B**) Only studies where type of platform was identified are shown. (**C**) Information extracted from WWF Terrestrial Ecoregions Map [42]. No drone studies were found in Protected Areas with Tropical and subtropical dry broadleaf forests.

The classification of the studies in categories that align with recurring aspects of conservation and management in protected areas [43] provides a framework that may help park-managers to identify feasible drone scenarios. The factors influencing effectiveness can be conveniently ascribed to the proposed categories and associated with consensual conservation actions [44]. In the next sections, we discuss the current state of the art and the challenges for the future integration of drones in protected areas.

3.1. State of the Art: Drones in Protected Areas

3.1.1. Wildlife Research and Management

Manned aircrafts have been traditionally used to complement ground-based wildlife surveys, but under-resourcing of many protected areas prevent their more widespread use. Besides, a significant number of aerial accidents with fatalities have been historically reported [45]. Moreover, aerial incursions are subject to visibility bias since a greater number of observers is required to guarantee an exhaustive count of populations and minimize errors [46]. Drones have emerged as a feasible alternative to surpass such inconveniences at small scales and complement modern wildlife conservation. Remotely sensed capabilities of drones offer a less invasive, non-hazardous, repetitive and reliable monitoring technique [47] to collect species abundance and distribution, document wildlife behavior, life-history and health status. Recent examples target terrestrial mammals [48–50]; marine mammals [51–55]; birds [11,56–60]; reptiles [15,61–64]; and fish [65,66]. Most surveys opted for both optical and thermal cameras, the latter especially appropriate to sense elusive species overnight, when the temperature differences between the animal body and the environment are greater [67]. Other studies implemented acoustic sensors to record songbirds [68] or combine drones with tracking systems aboard [9,69,70] to collect wildlife movement and environmental data. Researchers have also devised ways to use drones for insect monitoring [71], habitat modeling [72] and sampling [73].

Protected areas often face human-wildlife conflicts in populated areas bordering their limits [74]. Some studies described the use of drones in various management tasks, such as moving elephants out of human settlements [75], mapping wildlife damage on crops to calculate compensation costs [76] or dropping fake baits targeting feral species [77]. Drones constitute an attainable low-cost alternative to assess and reduce the risk that hazardous infrastructures [78,79] or mechanical harvesting [80,81] pose to wildlife. Lastly, fine-scale mapping of species distribution, land-use changes and water bodies using high resolution aerial imagery hold potential to complement epidemiological and zoonotic studies [82–85], and may serve as a rapid mechanism to inform prevention and reinforce biosecurity programs.

3.1.2. Ecosystem Monitoring

Protected areas are reference sites for ecological monitoring. These activities provide essential information to track ecosystem changes as a result of management and environmental factors [86]. Established methods for habitat monitoring range from in situ and airborne observations to satellite-based remote sensing. The latest generation of commercial satellite sensors [87] collect images at sub-meter resolution and entail remarkable technological advances to Earth observation, but the geographical availability of products is limited and not always rapidly available. Drones are particularly appropriated to timely survey small areas at unprecedented detail [88], could be adapted to carry sampling devices and take in-situ measurements [89], and may prove advantageous to monitor Essential Biodiversity Variables (EBVs) [90]. Similarly, mapping and quantifying ecosystem services with drones constitute an efficient means to inform site design and zoning, especially when the information available is scarce, outdated and based on coarse-resolution remote sensing images. Also, monitoring habitat degradation with drones in protected areas and borderlands [91,92] represents a novel method to assess the performance of conservation actions. Finally, fine-scale habitat assessment using high resolution maps could assist, selecting suitable reintroduction sites for endangered or locally extinct species [93].

Experimental drone monitoring projects have increased noticeably, both by governmental institutions [94] and research groups, for informing on the distribution [95], health [16,96], productivity [97], composition [98], structure [99,100] and biomass [101–104] of forests using both passive and active sensors [105]. As a consequence, drone applications for inventory, characterization and habitat restoration are maturing fast, but scaling-up and linking the collected information with that coming from satellite remote sensing remains a knowledge gap [106]. However, some studies represent a step in this direction, including the following: derive and enhance ground-based forest metrics to assist modeling of ecological process at regional scale [107], validate vegetation maps from drone image interpretation [108,109] or address the radiometric calibration of small multispectral cameras to allow comparisons with satellite data [110,111]. Drones have been used for community-based forest monitoring [112], and therefore suggested as an important asset to impulse the participation of developing countries in the carbon market (Reducing Emissions from

Deforestation and Forest Degradation, REDD) [113]. In addition, drones have operated successfully in different ecosystems to measure the spread of invasive species [114–118]; map coastal and marine habitats [119–125]; wetlands [126–130]; grasslands [17,131,132]; savannas [133,134]; glaciers [135–137]; polar areas [138,139]; and riparian ecosystems [140–143].

3.1.3. Law Enforcement

Efficient control and surveillance of illegal activities lead the ranking of measures for effective management of terrestrial [144] and marine [145] protected areas. These conservation actions aim to maintain the integrity of threatened species and ecosystems in the face of human pressures, but in practice suffer from serious deficiencies [146]. Enforcement is especially challenging in large protected areas where iconic species are on the verge of extinction due to illegal hunting, fishing, encroachment or habitat loss. Drones constitute a technological advance to complement insufficient staff and resourcing in anti-poaching [12,147–149] and other less contentious acts such as vandalism or bonfires in unauthorized areas [150,151]. Drone surveillance aim to autonomously detect and track subjects integrating live streaming visible and thermal camera systems with real time vision processing techniques. However, these applications are subjected to technological and legal constraints. Real-time recognition of suspicious activity or flying in adverse weather conditions remain a work in progress [152]. The relatively low maximum flight time of modest drones is a major obstacle to cover large areas [12], but progress is noticeable. Although the last generation of long-endurance fixed-wing and hybrid aerial platforms have higher autonomy, meeting the optimal specifications requires a considerable investment [153] with uncertain benefits, especially in developing countries [154]. Besides, the main barriers to protected areas surveillance using drones take place in the legislative and socio-political sphere. The flight rules often limit flying drones beyond the visual line of sight (BVLOS), above a certain altitude or at night, precluding the surveillance in periods of increased illegal activity. On the other hand, there are concerns about the alleged social and ethical implications of using drones with coercive purposes [155]. Duffy debated the advent of militarized conservation and stated that drones and similar technologies could contribute to human rights breaching [156], which may lessen the commitment of native communities [36,157] to protect their natural resources. Under these considerations, more research is needed to identify those technological advances and best practices that do not pose or minimize the risk to the privacy and welfare of people but serve for the purpose of surveillance. In this sense, thermal images reveal the temperature profile of the target, but lack the ability to collect sensitive personal information. Other measures can be taken to restrict the surveillance to previously defined zones and according to poaching threat maps [158] representing those areas with greatest pressures. In addition, some studies have remarked that the effectiveness of antipoaching depends on a greater allocation of resources [144]. For example, to improve the effectiveness of offshore guarding activities [159], patrol vessel could acquire waterproof rotary-wing or fixed-wing drones with float planes to persuade and record illegal fishing within the boundaries of marine protected areas. These evidences could be considered a reliable proof in court, even when offenders are seized outside the no-take zones [160]. Alternatively, there are some reported experiences where drones assisted counter-mapping with reasonable success [161,162]. With all due caution, these are some compelling reasons to encourage the development and implementation of drones to fight poaching. Nevertheless, the success of such initiatives might require a greater consensus among the parties involved and the development of multidisciplinary strategies that seek to solve these recurrent threats to biodiversity.

3.1.4. Ecotourism

Well-managed ecotourism promotes conservation and provides socioeconomic benefits to local communities. Otherwise, it may adversely affect the welfare of the animals and disrupt their habitats [163]. In the midst of the dilemma, drones have been proposed for recreational and educational purposes [164,165], document natural monuments and cultural sites [166]; and social research and visitor surveillance [167,168]. However, drone operations are susceptible to endanger wildlife [35], compromise tourist experience [153] or in case of accidents, lead to pollution or wildfires in sensitive areas due to the presence of toxic and flammable components. Subsequently, to restrain the uncontrolled presence of drones in protected areas, stakeholders agreed on a set of policies to establish permitted activities in Antarctica [169], opted for simpler rules and recommendations [170] or completely banned drones arguing safety reasons and wildlife impact [171]. Even when the economic benefits and leisure possibilities are promising, undesirable events and a lack of ethical practices could emphasize the negative connotations of drones to the detriment of their advantages. Thus, it would be advisable to be cautious in the face of a growing demand to incorporate drones into ecotourism services and continue working on a set of consensual measures to minimize the potential drawbacks drones may bring to protected areas.

3.1.5. Environmental Management and Disaster Response

Effectively managing protected areas requires continuous monitoring of environmental biophysical indicators to ensure that potential sources of contamination are controlled or below a safety threshold and, if necessary, take appropriate restoration measures. In many cases, a rapid response is crucial to diminish the effects that natural and man-made disasters pose to natural resources and human beings. Usually, these conservation actions combine fieldwork, airborne and satellite remote sensing. Drone capabilities provide a fine-scale alternative to remotely assist water, soil and air quality sampling [172–176], and enable rapid image acquisition to monitor erosion [177]; sediments dynamics [178,179]; forest windthrow [180]; habitat degradation [125]; landslides [181–183]; flood [184]; volcanic events [183,185,186]; oil spills [187]; and wildfires [188–190] at different stages. Drones may also serve as valuable tools for rangers in search and rescue missions in marine and remote mountainous regions [191,192]. Besides, there are a variety of plausible scenarios where drones can prove to be useful, such as detecting marine litter [193–195], inspect facilities [196]; collect information gathered from environmental sensor networks [197]; or support plant invasion monitoring [198] and control by means of aerially deployed herbicide on targeted species [199].

3.2. Current Challenges on the Integration of Drones in Protected Areas

3.2.1. Legal Barriers and Ethical Constraints

Drone operations face important social and legal barriers that undermine their potential in the civilian sphere [36,200,201]. Not without founded reasons, an overly restrictive and indiscriminate regulatory framework arguing privacy and safety issues is currently limiting the applications of drones in the field of conservation. This highlights the urgent need to seek consensus among countries and adapt legislation to distinguish between the purpose of leisure, research and management [202].

3.2.2. Impact of Drones on Wildlife and Ecosystems

Animal welfare and alteration of sensitive habitat in wildlife management and ecological research is a source of strong debate [203,204]. Some authors have reported disturbance effects of drones on birds [57,205–209], reptiles [210] and mammals [211–213]. Despite a greater degree of awareness reflected in a emergent set of guidelines to minimize the impacts on wildlife [35,56,214,215], most studies marginally inform reactions and further trials aimed at quantifying changes in behavioral patterns and physiological effects targeting a broader group of species is recommended. An optimal trade-off between benefits and environmental costs should be weighed [216,217]. By designing quieter, non-polluting and safer components, along with following up the suggested flight patterns, the impact on wildlife and ecosystems could be reduced and its objective and unbiased observation facilitated [47,204]. Therefore, drones have great potential to evolve, replacing more invasive monitoring techniques. This should be consciously considered by those reluctant to integrate drones in research and conservation activities. Step by step, a code of best practice and recommendations could be continuously updated based on lessons learned [206], forming the basis for wildlife certified drone operators [35].

3.2.3. Costs of Drone Operation

Expenses derived from using drones in the long term are difficult to quantify [218] and depend on a confluence of factors. Some of the applications described above rely on the acquisition of sophisticated on-board instruments, devices and sensors, advanced communications system or gas-powered engines for longer endurance and heavier payloads. The large volume of data collected must be conveniently stored and processed, which often require qualified staff and adequate IT (Information Technology) infrastructures. In addition, operations with drones are not exempt from accidents, which may compromise the viability of some projects. The payload is usually the most expensive part of the platform, and this often breaks down. Park managers should be aware that there is not a single solution covering all the conservation purposes [219] and a trade-off analysis among available platforms and sensors should be pondered. In this regard, do-it-yourself (DIY) drones can be equipped with a flexible array of sensors and according to very specific requirements, but extra time and experience is required for the correct assemblage and configuration of parts. Since ready to fly commercial platforms are tested and proven systems, it could be argued that they present more reliable capabilities than custom-built drones. Moreover, the consumer market shows a gradual drop on prices in higher performance platforms [220]. Suppliers often provide support, training and companion software, albeit services could be occasionally charged. Nonetheless, there is general agreement that costs associated with drones are lower compared to established methods (Table 3), such as manned aircraft and ground incursions [13,178,211], at least for mapping small and medium scale areas. Although the benefits of monitoring greater extensions with drones remain challenging according to the state of the art, the situation is likely to be more favorable with the advent of more efficient aerial platforms.

Table 3. Examples of studies reporting favorable use of drones compared with established methods.

Study	Aim	Established Methods	Using Drones
[173]	Water Sampling	**Boat sampling** • 3 scientists, 1 boat, 1 truck, 1 trailer. • Slow, spatially restricted. • Expensive and laborious deployment • lake sampled/10–15-h day.	• 2 h, 1 scientist, 1 drone. • Sample all lakes at very high spatio-temporal resolution.
[57]	Nesting status of birds	**Climbing trees** • 2500$ • 2 people and climbing gear. • 33 min/inspection	• 1000$ • 1 person and drone • 4:30 min/inspection
[57]	Elasmobranchs densities	**Fishing methods, diver surveys, video cameras, aerial surveys** • Potential invasive methods • Prohibitive cost. • Risk for observers and observer bias.	• <2500 $ • Short period of time. • True densities
[61]	Crocodile nesting behavior	**Helicopter, airboat, ground surveys** • Prohibitive cost. • Dangerous incursions.	• Low cost, repeatability, and flexibility
[221]	Mangrove forest inventory	**Fieldwork** • Laborious and costly • Trade-off sample size and frequency • Located in remote areas. • Disturbance of fauna and flora	• Consumer-grade drone 1200 $ • Above ground biomass estimation. • Increase sampling frequency • Less invasive

3.2.4. Technological Challenges

As previously noted, the massive volume of data that sensors collect in the course of the surveys need to be stored, processed and analyzed, causing severe procedural bottlenecks [6] that need to be solved. When using aerial images for wildlife census, the manual counting and identification of individuals represent a considerable investment in time and costs. Progress in computer vision and machine learning are intended to automate such routine tasks [52,81,222–228]. Despite encouraging results [229], these methods are only available for relatively easy to spot species in open natural environments and require highly qualified personnel to offer reliable results. In addition, further research is required to assess the overall performance of drone data collection techniques to address the analysis and modelling of species distribution, especially in comparison with more mature statistical and sampling methods [58]. On the other hand, traditional pixel-based algorithms are rather inefficient when processing very high resolution images [128]. Therefore, object-based image analysis (OBIA) and deep learning techniques [230] will likely prevail during the next generation of land-cover, habitat and vegetation classification methods [8]. The arrival of affordable hyperspectral miniaturized sensors [124,128,231,232], will bring more complexity to the matter, requiring novel analytical approaches not currently implemented. Conversely, the entire photogrammetric process is well documented [233] and supported by commercial desktop and mobile applications, but also open source solutions [21], probably at expense of a major level of expertise [234,235]. Drones using Real Time Kinematic (RTK) and Post Processing Kinematic (PPK) techniques can produce survey-grade maps without requiring labor intensive ground control points (GCPs). Yet, the radiometric calibration of aerial images requires additional improvements [37] since it is considered a crucial step to carry out multi-temporal studies [236]. The confluence of big data [237], networked drones [238,239], artificial intelligence and sensors will bring new unforeseen perspectives to conservation, but integration of products and services to deliver off-the-shelf management solutions are still in their infancy.

3.3. Linking Drone Platforms and Sensors with Conservation

Park managers considering the acquisition of drones may need expert guidance to select the most suitable platform and sensor for each purpose. Here we provide a brief summary of most common imaging and ranging sensors (Table 4). Consumer grade cameras are adequate for general mapping and photogrammetric tasks. Sensor size, focal length and lens quality are the main camera factors that influence the accuracy of the survey. More advance remote sensing applications require the adoption of multispectral and hyperspectral sensors. The former encompasses both modified RGB cameras to near infrared and multispectral cameras with great prospects for precision agriculture, forestry and a broad range of vegetation studies [240]. Hyperspectral sensors collect information in multiple bands across the electromagnetic spectrum, and are of great interest to remotely observe the spectral response of many distinct biophysical parameters [22] and physiological process of organisms [124]. These families of sensors require radiometric calibration to account for variable lighting conditions and retrieve physical quantities that can be compared in time and with other sensors [241]. Thermal infrared cameras can remotely sense heat even in low visibility conditions and are ordinarily used for industrial inspection and surveillance, but also in soil science [242] and animal ecology [64]. Thermal sensitivity, expressed as the ability of the sensor to discriminate differences of temperatures even in low contrast scenes, is one of the most important technical aspects to increase the detection rate of wildlife [52]. LiDAR instruments are relatively expensive active sensors that can penetrate the canopy and derive accurate three-dimensional forest metrics and terrain models. However, structure-from-motion (SfM) [243] imaging techniques based on standard RGB cameras represent a low-cost alternative with limited, but reasonable results. In terms of platforms, long-endurance fixed-wing drones are preferred when surveying large areas and when landing is not a problem. Conversely, rotary-wing platforms are more versatile, and can operate in a diverse range of situations where precise flights prove more advantageous, such as in confined spaces and close-range inspection tasks, marine settings and terrestrial areas with steep terrain, or extensive vegetation cover.

Table 4. Suitable sensors for research and management tasks.

Sensor	Applications
Visible RGB	Aerial photography, habitat mapping, photogrammetry, 3D Modeling, inspection, wildlife surveys (identification), landslides
Multispectral	Vegetation indices, productivity, water quality, geological surveys
Hyperspectral	Vegetation studies, biophysical variables, ecological processes, forest health, chlorophyll content, insect outbreaks.
Thermal	Inspection, wildlife surveys (detection), surveillance, wildfires, soil temperature, volcanology
LiDAR	3D Modeling, topographical maps, forest inventory and metrics (structure, biomass, tree volume, canopy height, leaf area index)

3.4. Knowledge Gaps and Recommendations for Future Research

The variety of information gathered from drones represents a great opportunity to complement ongoing Earth Observation programs aimed to monitor anthropogenic pressures threatening the ecological integrity of protected areas [244]. Drones can be rapidly deployed there, where early sign of disturbance have been previously detected using satellite images and environmental sensor networks [245]. Although many protected areas are too large to be mapped using drones, there are small, inaccessible and environmentally sensitive terrestrial and marine areas (ESAs) with important ecological values that could take advantage from drones. Once the use of drones has proven feasible in many different fields of application, it would be of interest that research focuses on methods to produce a set of ecological indicators in line with established monitoring frameworks [246]. For example, a wide range of biodiversity metrics, ecosystem processes and natural and anthropogenic stressors could be measured or derived, but further efforts are required to transfer advances on the field into accessible products for direct use at management levels. Table 5 suggest some potential challenges that can help to guide future research in the field.

Table 5. Challenges for the effective implementation of drones in protected areas.

Management Categories	Challenges
Wildlife Research and Management	• Development of drones to minimize impact of wildlife. • Optimization of automatic pattern recognition algorithms. • Robust sampling design/limited statistical power. • Integrating movement and visible/thermal data. • Population structure and function, wildlife traits.
Ecosystem Monitoring	• Consistent ecological indicators. • Multitemporal studies. • Targeting Essential Biodiversity Variables (EBVs). • Multiscale studies/linking drones with Earth Observation systems. • Mapping of aquatic environments/bathymetry maps • Machine learning methods (neural networks, etc.) • Ecosystem services/area designation and performance. • Habitat suitability/species reintroduction studies
Law Enforcement	• Research required to assess the performance of drones to reduce illegal activities. • Test hybrid (VTOL) platforms. • Marine Protected Areas: Drones/Vessel patrols • Focus on poaching, but there are other important human intrusions in protected areas that could benefit from drones (illegal logging, mining, etc.) • Threat maps.
Ecotourism	• Cost/benefit analysis • Potential to introduce virtual flights. • Fine-scale geofencing maps (Detailed map of sites where drone flights are allowed/conditioned/restricted)
Environmental Management and Disaster Response	• Move from prototypes to products and services. • Implementation of Regional/Global Infrastructures for decision support. • Satellite/Drone Remote Sensing integrative approach to model disturbance regimes.

Supplementary Materials: The following are available online at http://www.mdpi.com/2504-446X/3/1/10/s1, **Figure S1.** PRISMA Flowchart; **Table S1**. List of studies.

Author Contributions: J.J.L. and M.M.P conceived and designed the study; J.J.L. collected and analyzed the studies; J.J.L. and M.M.P wrote the manuscript.

Funding: This research received no external funding.

Acknowledgments: The authors would like to thank the Editorial Office, and anonymous reviewers for their valuable comments and suggestions.

Conflicts of Interest: The authors declare no conflict of interest.

References

1. Dudley, N. *Guidelines for Protected Area Management Categories*; IUCN: Gland, Switzerland, 2008; Volume 3, ISBN 978-2-8317-1086-0.
2. Watson, J.E.M.; Dudley, N.; Segan, D.B.; Hockings, M. The performance and potential of protected areas. *Nature* **2014**, *515*, 67–73. [CrossRef] [PubMed]
3. Juffe-Bignoli, D.; Burgess, N.D.; Bingham, H.; Belle, E.M.S.; de Lima, M.G.; Deguignet, M.; Bertzky, B.; Milam, N.; Martinez-Lopez, J.; Lewis, E.; et al. *Protected Planet Report 2014*; UNEP-WCMC: Cambridge, UK, 2014; ISBN 9789280734164.
4. Gonçalves, J.; Henriques, R.; Alves, P.; Sousa-Silva, R.; Monteiro, A.T.; Lomba, Â.; Marcos, B.; Honrado, J. Evaluating an unmanned aerial vehicle-based approach for assessing habitat extent and condition in fine-scale early successional mountain mosaics. *Appl. Veg. Sci.* **2016**, *19*, 132–146. [CrossRef]
5. Lopoukhine, N.; Crawhall, N.; Dudley, N.; Figgis, P.; Karibuhoye, C.; Laffoley, D.; Londoño, J.M.; MacKinnon, K.; Sandwith, T. Protected areas: providing natural solutions to 21st Century challenges. *Surveys and Perspectives Integrating Environment and Society* **2012**, *5*, 1–16. [CrossRef]
6. Pimm, S.L.; Alibhai, S.; Bergl, R.; Dehgan, A.; Giri, C.; Jewell, Z.; Joppa, L.; Kays, R.; Loarie, S. Emerging technologies to conserve biodiversity. *Trends Ecol. Evol.* **2015**, *30*, 685–696. [CrossRef] [PubMed]
7. Melesse, A.; Weng, Q.; Prasad, S.; Senay, G. Remote Sensing Sensors and applications in environmental resources mapping and modelling. *Sensors* **2007**, *7*, 3209–3241. [CrossRef] [PubMed]
8. Whitehead, K.; Hugenholtz, C.H. Remote sensing of the environment with small unmanned aircraft systems (UASs), part 1: A review of progress and challenges. *J. Unmanned Veh. Syst.* **2014**, *2*, 69–85. [CrossRef]
9. Rodríguez, A.; Negro, J.J.; Mulero, M.; Rodríguez, C.; Hernández-Pliego, J.; Bustamante, J. The Eye in the Sky: Combined Use of Unmanned Aerial Systems and GPS Data Loggers for Ecological Research and Conservation of Small Birds. *PLoS One* **2012**, *7*. [CrossRef]
10. Linchant, J.; Lisein, J.; Semeki, J.; Lejeune, P.; Vermeulen, C. Are unmanned aircraft systems (UASs) the future of wildlife monitoring? A review of accomplishments and challenges. *Mamm. Rev.* **2015**, *45*, 239–252. [CrossRef]
11. Christie, K.S.; Gilbert, S.L.; Brown, C.L.; Hatfield, M.; Hanson, L. Unmanned aircraft systems in wildlife research: Current and future applications of a transformative technology. *Front. Ecol. Environ.* **2016**, *14*, 241–251. [CrossRef]
12. Mulero-Pázmány, M.; Stolper, R.; Van Essen, L.D.; Negro, J.J.; Sassen, T. Remotely piloted aircraft systems as a rhinoceros anti-poaching tool in Africa. *PLoS One* **2014**, *9*, 1–10. [CrossRef]
13. Koh, L.P.; Wich, S.A. Dawn of drone ecology: low-cost autonomous aerial vehicles for conservation. *Trop. Conserv. Sci.* **2012**, *5*, 121–132. [CrossRef]
14. Chabot, D.; Bird, D.M. Wildlife research and management methods in the 21st century: Where do unmanned aircraft fit in? *J. Unmanned Veh. Syst.* **2015**, *3*, 137–155. [CrossRef]
15. Schofield, G.; Katselidis, K.A.; Lilley, M.K.S.; Reina, R.D.; Hays, G.C. Detecting elusive aspects of wildlife ecology using drones: new insights on the mating dynamics and operational sex ratios of sea turtles. *Funct. Ecol.* **2017**, *38*, 42–49. [CrossRef]
16. Näsi, R.; Honkavaara, E.; Lyytikäinen-Saarenmaa, P.; Blomqvist, M.; Litkey, P.; Hakala, T.; Viljanen, N.; Kantola, T.; Tanhuanpää, T.; Holopainen, M. Using UAV-based photogrammetry and hyperspectral imaging for mapping bark beetle damage at tree-level. *Remote Sens.* **2015**, *7*, 15467–15493. [CrossRef]

17. Wang, D.; Xin, X.; Shao, Q.; Brolly, M.; Zhu, Z.; Chen, J. Modeling aboveground biomass in Hulunber grassland ecosystem by using unmanned aerial vehicle discrete lidar. *Sensors* **2017**, *17*, 180. [CrossRef] [PubMed]
18. Villa, T.; Gonzalez, F.; Miljievic, B.; Ristovski, Z.; Morawska, L. An overview of small unmanned aerial vehicles for air quality measurements: Present applications and future prospectives. *Sensors* **2016**, *16*, 1072. [CrossRef] [PubMed]
19. Pádua, L.; Vanko, J.; Hruška, J.; Adão, T.; Sousa, J.J.; Peres, E.; Morais, R. UAS, sensors, and data processing in agroforestry: a review towards practical applications. *Int. J. Remote Sens.* **2017**, *38*, 2349–2391. [CrossRef]
20. Hassanalian, M.; Abdelkefi, A. Classifications, applications, and design challenges of drones: A review. *Prog. Aerosp. Sci.* **2017**, *91*, 99–131. [CrossRef]
21. Colomina, I.; Molina, P. Unmanned aerial systems for photogrammetry and remote sensing: A review. *ISPRS J. Photogramm. Remote Sens.* **2014**, *92*, 79–97. [CrossRef]
22. Adão, T.; Hruška, J.; Pádua, L.; Bessa, J.; Peres, E.; Morais, R.; Sousa, J. Hyperspectral imaging: A review on uav-based sensors, data processing and applications for agriculture and forestry. *Remote Sens.* **2017**, *9*, 1110. [CrossRef]
23. Brown, A. Grand Canyon Park Rangers Use Drones to Search for Missing Hikers. Available online: http://www.thedrive.com/aerial/9619/grand-canyon-park-rangers-use-drones-to-search-for-missing-hikers?# (accessed on 27 September 2018).
24. Margaritoff, M. Australian Rangers Use Drones to Monitor and Protect Endangered Turtles From Predators. Available online: http://www.thedrive.com/tech/22747/australian-rangers-use-drones-to-monitor-and-protect-endangered-turtles-from-predators (accessed on 26 September 2018).
25. Richmond, B.C. Drones, dogs and DNA the latest weapons against invasive species. Available online: https://www.theglobeandmail.com/news/national/drones-dogs-and-dna-the-latest-weapons-against-invasive-species/article28531824/ (accessed on 27 September 2018).
26. Beaubien, J. Eyes In The Sky: Foam Drones Keep Watch On Rain Forest Trees. Available online: https://www.npr.org/sections/goatsandsoda/2015/05/19/398765759/eyes-in-the-sky-styrofoam-drones-keep-watch-on-rainforest-trees?t=1540327094055 (accessed on 28 September 2018).
27. Crooks, J. Drone Collects Information to Benefit Great Lakes. Available online: https://www.fs.fed.us/blogs/drone-collects-information-benefit-great-lakes (accessed on 27 September 2018).
28. Nuwer, R. High Above, Drones Keep Watchful Eyes on Wildlife in Africa. Available online: https://www.nytimes.com/2017/03/13/science/drones-africa-poachers-wildlife.html (accessed on 27 September 2018).
29. Kohler, P. Launching the Marine Litter DRONET. Available online: https://www.theplastictide.com/blog-1/2018/4/22/launching-the-marine-litter-dronet (accessed on 22 September 2018).
30. Cirino, E. Drones Help Find Massive Penguin Colonies Hiding in Plain Sight. Available online: https://www.newsdeeply.com/oceans/articles/2018/03/05/drones-help-find-massive-penguin-colonies-hiding-in-plain-sight (accessed on 18 September 2018).
31. DRONE HOPPER, S.L. Drone Hopper | Firefigthing Drone. Available online: https://www.drone-hopper.com/home (accessed on 2 September 2018).
32. Apprill, A.; Miller, C.A.; Moore, M.J.; Durban, J.W.; Fearnbach, H.; Barrett-Lennard, L.G. Extensive Core Microbiome in Drone-Captured Whale Blow Supports a Framework for Health Monitoring. *mSystems* **2017**. [CrossRef] [PubMed]
33. Foxx, C. Drones scatter mosquitoes to fight diseases. Available online: https://www.bbc.co.uk/news/technology-42066518 (accessed on 21 September 2018).
34. Stone, E. Drones Spray Tree Seeds From the Sky to Fight Deforestation. Available online: https://news.nationalgeographic.com/2017/11/drones-plant-trees-deforestation-environment/?user.testname=none (accessed on 21 September 2018).
35. Mulero-Pázmány, M.; Jenni-Eiermann, S.; Strebel, N.; Sattler, T.; Negro, J.J.; Tablado, Z. Unmanned aircraft systems as a new source of disturbance for wildlife: A systematic review. *PLoS One* **2017**, *12*, e0178448. [CrossRef] [PubMed]
36. Sandbrook, C. The social implications of using drones for biodiversity conservation. *Ambio* **2015**, *44*, 636–647. [CrossRef]

37. Manfreda, S.; McCabe, M.F.; Miller, P.E.; Lucas, R.; Madrigal, V.P.; Mallinis, G.; Ben Dor, E.; Helman, D.; Estes, L.; Ciraolo, G.; et al. On the use of unmanned aerial systems for environmental monitoring. *Remote Sens.* **2018**, *10*. [CrossRef]

38. Bruner, A.G. Effectiveness of parks in protecting tropical biodiversity. *Science* **2001**, *291*, 125–128. [CrossRef] [PubMed]

39. Leverington, F.; Costa, K.L.; Courrau, J.; Pavese, H.; Nolte, C.; Marr, M.; Coad, L.; Burgess, N.; Bomhard, B.; Hockings, M.; et al. Management effectiveness evaluation in protected areas—A global study. Second edition 2010. *Environ. Manage.* **2010**, *46*, 685–698. [CrossRef]

40. Avramovic Danijela Evaluation of protected area management effectiveness—An overview of methodologies. *Saf. Eng.* **2016**, *6*, 29–35. [CrossRef]

41. UNEP-WCMC; IUCN Protected Planet: The World Database on Protected Areas (WDPA)/The Global Database on Protected Areas Management Effectiveness (GD-PAME). Available online: https://www.protectedplanet.net (accessed on 22 October 2018).

42. Olson, D.M.; Dinerstein, E.; Wikramanayake, E.D.; Burgess, N.D.; Powell, G.V.N.; Underwood, E.C.; D'amico, J.A.; Itoua, I.; Strand, H.E.; Morrison, J.C.; et al. Terrestrial Ecoregions of the World: A New Map of Life on Earth. *Bioscience* **2001**. [CrossRef]

43. Stoll-Kleemann, S. Evaluation of management effectiveness in protected areas: Methodologies and results. *Basic Appl. Ecol.* **2010**, *11*, 377–382. [CrossRef]

44. Salafsky, N.; Salzer, D.; Stattersfield, A.J.; Hilton-Taylor, C.; Neugarten, R.; Butchart, S.H.M.; Collen, B.; Cox, N.; Master, L.L.; O'Connor, S.; et al. A standard lexicon for biodiversity conservation: Unified classifications of threats and actions. *Conserv. Biol.* **2008**, *22*, 897–911. [CrossRef]

45. Sasse, D.B. Job-related mortality of wildlife workers in the United States, 1937-2000. *Wildl. Soc. Bull.* **2003**, *31*, 1000–1003.

46. Lubow, B.C.; Ransom, J.I. Practical bias correction in aerial surveys of large mammals: Validation of hybrid double-observer with sightability method against known abundance of feral horse (Equus caballus) populations. *PLoS One* **2016**. [CrossRef] [PubMed]

47. Jewell, Z. Effect of Monitoring Technique on Quality of Conservation Science. *Conserv. Biol.* **2013**, *27*, 501–508. [CrossRef] [PubMed]

48. Chrétien, L.-P.; Théau, J.; Ménard, P. Visible and thermal infrared remote sensing for the detection of white-tailed deer using an unmanned aerial system. *Wildl. Soc. Bull.* **2016**, *40*, 181–191. [CrossRef]

49. Wich, S.; Dellatore, D.; Houghton, M.; Ardi, R.; Koh, L.P. A preliminary assessment of using conservation drones for Sumatran orangutan (Pongo abelii) distribution and density. *J. Unmanned Veh. Syst.* **2016**, *4*, 45–52. [CrossRef]

50. Stark, D.J.; Vaughan, I.P.; Evans, L.J.; Kler, H.; Goossens, B. Combining drones and satellite tracking as an effective tool for informing policy change in riparian habitats: A proboscis monkey case study. *Remote Sens. Ecol. Conserv.* **2017**, 1–9. [CrossRef]

51. Sweeney, K.L.; Helker, V.T.; Perryman, W.L.; LeRoi, D.J.; Fritz, L.W.; Gelatt, T.S.; Angliss, R.P. Flying beneath the clouds at the edge of the world: using a hexacopter to supplement abundance surveys of Steller sea lions (Eumetopias jubatus) in Alaska. *J. Unmanned Veh. Syst.* **2016**, *4*, 70–81. [CrossRef]

52. Seymour, A.C.; Dale, J.; Hammill, M.; Halpin, P.N.; Johnston, D.W. Automated detection and enumeration of marine wildlife using unmanned aircraft systems (UAS) and thermal imagery. *Sci. Rep.* **2017**, *7*, 45127. [CrossRef]

53. Hodgson, A.; Peel, D.; Kelly, N. Unmanned aerial vehicles for surveying marine fauna: assessing detection probability. *Ecol. Appl.* **2017**, *27*, 1253–1267. [CrossRef]

54. Torres, L.G.; Nieukirk, S.L.; Lemos, L.; Chandler, T.E. Drone Up! Quantifying Whale Behavior From a New Perspective Improves Observational Capacity. *Front. Mar. Sci.* **2018**, *5*, 1–14. [CrossRef]

55. Pirotta, V.; Smith, A.; Ostrowski, M.; Russell, D.; Jonsen, I.D.; Grech, A.; Harcourt, R. An Economical Custom-Built Drone for Assessing Whale Health. *Front. Mar. Sci.* **2017**, *4*, 1–12. [CrossRef]

56. Junda, J.; Greene, E.; Bird, D.M. Proper flight technique for using a small rotary-winged drone aircraft to safely, quickly, and accurately survey raptor nests. *J. Unmanned Veh. Syst.* **2015**, *3*, 222–236. [CrossRef]

57. Weissensteiner, M.H.; Poelstra, J.W.; Wolf, J.B.W. Low-budget ready-to-fly unmanned aerial vehicles: An effective tool for evaluating the nesting status of canopy-breeding bird species. *J. Avian Biol.* **2015**, *46*, 425–430. [CrossRef]

58. Hodgson, J.C.; Baylis, S.M.; Mott, R.; Herrod, A.; Clarke, R.H. Precision wildlife monitoring using unmanned aerial vehicles. *Sci. Rep.* **2016**, *6*, 22574. [CrossRef] [PubMed]

59. Sardà-Palomera, F.; Bota, G.; Padilla, N.; Brotons, L.; Sardà, F. Unmanned aircraft systems to unravel spatial and temporal factors affecting dynamics of colony formation and nesting success in birds. *J. Avian Biol.* **2017**. [CrossRef]

60. Han, Y.G.; Yoo, S.H.; Kwon, O. Possibility of applying unmanned aerial vehicle (UAV) and mapping software for the monitoring of waterbirds and their habitats. *J. Ecol. Environ.* **2017**, *41*, 1–7. [CrossRef]

61. Evans, L.J.; Jones, T.H.; Pang, K.; Evans, M.N.; Saimin, S.; Goossens, B. Use of drone technology as a tool for behavioral research: A case study of crocodilian nesting. *Herpetol. Conserv. Biol.* **2015**, *10*, 90–98. [CrossRef]

62. Elsey, R.M.; Trosclair, P.L., III. the use of an unmanned aerial vehicle to locate alligator nests. *Southeast. Nat.* **2016**, *15*, 76–82. [CrossRef]

63. Sykora-Bodie, S.T.; Bezy, V.; Johnston, D.W.; Newton, E.; Lohmann, K.J. Quantifying Nearshore Sea Turtle Densities: Applications of Unmanned Aerial Systems for Population Assessments. *Sci. Rep.* **2017**, *7*, 1–7. [CrossRef]

64. Rees, A.; Avens, L.; Ballorain, K.; Bevan, E.; Broderick, A.; Carthy, R.; Christianen, M.; Duclos, G.; Heithaus, M.; Johnston, D.; et al. The potential of unmanned aerial systems for sea turtle research and conservation: a review and future directions. *Endanger. Species Res.* **2018**, *35*, 81–100. [CrossRef]

65. Groves, P.A.; Alcorn, B.; Wiest, M.M.; Maselko, J.M.; Connor, W.P. Testing unmanned aircraft systems for salmon spawning surveys. *Facets* **2016**, *1*, 187–204. [CrossRef]

66. Kiszka, J.J.; Mourier, J.; Gastrich, K.; Heithaus, M.R. Using unmanned aerial vehicles (UAVs) to investigate shark and ray densities in a shallow coral lagoon. *Mar. Ecol. Prog. Ser.* **2016**, *560*, 237–242. [CrossRef]

67. Kellenberger, B.; Marcos, D.; Tuia, D. Detecting mammals in UAV images: Best practices to address a substantially imbalanced dataset with deep learning. *Remote Sens. Environ.* **2018**, *216*, 139–153. [CrossRef]

68. Wilson, A.M.; Barr, J.; Zagorski, M. The feasibility of counting songbirds using unmanned aerial vehicles. *Auk* **2017**, *134*, 350–362. [CrossRef]

69. Mulero-Pázmány, M.; Barasona, J.Á.; Acevedo, P.; Vicente, J.; Negro, J.J. Unmanned Aircraft Systems complement biologging in spatial ecology studies. *Ecol. Evol.* **2015**, *5*, 4808–4818. [CrossRef] [PubMed]

70. Tremblay, J.A.; Desrochers, A.; Aubry, Y.; Pace, P.; Bird, D.M. A Low-Cost Technique for Radio-Tracking Wildlife Using a Small Standard Unmanned Aerial Vehicle. *J. Unmanned Veh. Syst.* **2016**, juvs-2016-0021. [CrossRef]

71. Ivosevic, B.; Han, Y.-G.; Kwon, O. Monitoring butterflies with an unmanned aerial vehicle: current possibilities and future potentials. *J. Ecol. Environ.* **2017**, *41*, 12. [CrossRef]

72. Habel, J.C.; Teucher, M.; Ulrich, W.; Bauer, M.; Rödder, D. Drones for butterfly conservation: larval habitat assessment with an unmanned aerial vehicle. *Landsc. Ecol.* **2016**, *31*, 2385–2395. [CrossRef]

73. Kim, H.G.; Park, J.-S.; Lee, D.-H. Potential of Unmanned Aerial Sampling for Monitoring Insect Populations in Rice Fields. *Florida Entomol.* **2018**, *101*, 330–334. [CrossRef]

74. Ogra, M.V. Human-wildlife conflict and gender in protected area borderlands: A case study of costs, perceptions, and vulnerabilities from Uttarakhand (Uttaranchal), India. *Geoforum* **2008**. [CrossRef]

75. Hahn, N.; Mwakatobe, A.; Konuche, J.; de Souza, N.; Keyyu, J.; Goss, M.; Chang'a, A.; Palminteri, S.; Dinerstein, E.; Olson, D. Unmanned aerial vehicles mitigate human–elephant conflict on the borders of Tanzanian Parks: a case study. *Oryx* **2017**, *51*, 513–516. [CrossRef]

76. Michez, A.; Morelle, K.; Lehaire, F.; Widar, J.; Authelet, M.; Vermeulen, C.; Lejeune, P.; Michez, A.; Morelle, K.; Lehaire, F.; et al. Use of unmanned aerial system to assess wildlife (Sus scrofa) damage to crops (Zea mays). *J. Unmanned Veh. Sys* **2016**, *4*, 266–275. [CrossRef]

77. Johnston, M.; McCaldin, G.; Rieker, A. Assessing the availability of aerially-delivered baits to feral cats through rainforest canopy using unmanned aircraft. **2016**. [CrossRef]

78. Mulero-Pázmány, M.; Negro, J.J.; Ferrer, M. A low cost way for assessing bird risk hazards in power lines: Fixed-wing small unmanned aircraft systems. *J. Unmanned Veh. Syst.* **2014**, *02*, 5–15. [CrossRef]

79. Lobermeier, S.; Moldenhauer, M.; Peter, C.; Slominski, L.; Tedesco, R.; Meer, M.; Dwyer, J.; Harness, R.; Stewart, A. Mitigating avian collision with power lines: a proof of concept for installation of line markers via unmanned aerial vehicle. *J. Unmanned Veh. Syst.* **2015**, *3*, 252–258. [CrossRef]

80. Israel, M. A UAV-based roe deer fawm detection system. *Int. Arch. Photogramm. Remote Sens. Spat. Inf. Sci.* **2012**, *38*, 1–5. [CrossRef]

81. Christiansen, P.; Steen, K.A.; Jørgensen, R.N.; Karstoft, H. Automated detection and recognition of wildlife using thermal cameras. *Sensors* **2014**, *14*, 13778–13793. [CrossRef]

82. Fornace, K.M.; Drakeley, C.J.; William, T.; Espino, F.; Cox, J. Mapping infectious disease landscapes: Unmanned aerial vehicles and epidemiology. *Trends Parasitol.* **2014**, *30*, 514–519. [CrossRef]

83. Barasona, J.A.; Mulero-Pázmány, M.; Acevedo, P.; Negro, J.J.; Torres, M.J.; Gortázar, C.; Vicente, J. Unmanned aircraft systems for studying spatial abundance of ungulates: Relevance to spatial epidemiology. *PLoS One* **2014**, *9*, 1–17. [CrossRef]

84. Hardy, A.; Makame, M.; Cross, D.; Majambere, S.; Msellem, M. Using low-cost drones to map malaria vector habitats. *Parasit. Vectors* **2017**, *10*, 29. [CrossRef]

85. Laguna, E.; Barasona, J.A.; Triguero-Ocaña, R.; Mulero-Pázmány, M.; Negro, J.J.; Vicente, J.; Acevedo, P. The relevance of host overcrowding in wildlife epidemiology: A new spatially explicit aggregation index. *Ecol. Indic.* **2018**, *84*, 695–700. [CrossRef]

86. Gibbs, J.P.; Snell, H.L.; Causton, C.E. Effective monitoring for adaptive wildlife management: Lessons from the Galapagos Islands. *J. Wildl. Manage.* **1999**, *63*, 1055–1065. [CrossRef]

87. Neumann, W.; Martinuzzi, S.; Estes, A.B.; Pidgeon, A.M.; Dettki, H.; Ericsson, G.; Radeloff, V.C. Opportunities for the application of advanced remotely-sensed data in ecological studies of terrestrial animal movement. *Mov. Ecol.* **2015**. [CrossRef] [PubMed]

88. Whitehead, K.; Hugenholtz, C. Remote Sensing of the Environment with Small Unmanned Aircraft Systems (UASs), Part 2: scientific and commercial applications. *J. Unmanned Veh. Syst.* **2014**, *102*, 86–102. [CrossRef]

89. Gross, J.E.; Goetz, S.J.; Cihlar, J. Application of remote sensing to parks and protected area monitoring: Introduction to the special issue. *Remote Sens. Environ.* **2009**, *113*, 1343–1345. [CrossRef]

90. Vihervaara, P.; Auvinen, A.P.; Mononen, L.; Törmä, M.; Ahlroth, P.; Anttila, S.; Böttcher, K.; Forsius, M.; Heino, J.; Heliölä, J.; et al. How Essential Biodiversity Variables and remote sensing can help national biodiversity monitoring. *Glob. Ecol. Conserv.* **2017**, *10*, 43–59. [CrossRef]

91. Mas, J.F. Assessing protected area effectiveness using surrounding (buffer) areas environmentally similar to the target area. *Environ. Monit. Assess.* **2005**, *105*, 69–80. [CrossRef]

92. Ewers, R.M.; Rodrigues, A.S.L. Estimates of reserve effectiveness are confounded by leakage. *Trends Ecol. Evol.* **2008**, *23*, 113–116. [CrossRef]

93. Freemantle, T.P.; Wacher, T.; Newby, J.; Pettorelli, N. Earth observation: Overlooked potential to support species reintroduction programmes. *Afr. J. Ecol.* **2013**. [CrossRef]

94. Cress, J.; Hutt, M.; Sloan, J.; Bauer, M.; Feller, M.; Goplen, S. *US Geological Survey Unmanned Aircraft Systems (UAS) Roadmap 2014*; USGS: Reston, VA, USA, 2015.

95. Puliti, S.; Olerka, H.; Gobakken, T.; Næsset, E. Inventory of Small Forest Areas Using an Unmanned Aerial System. *Remote Sens.* **2015**, *7*, 9632–9654. [CrossRef]

96. Michez, A.; Piégay, H.; Lisein, J.; Claessens, H.; Lejeune, P. Classification of riparian forest species and health condition using multi-temporal and hyperspatial imagery from unmanned aerial system. *Environ. Monit. Assess.* **2016**, *188*, 1–19. [CrossRef]

97. Tian, J.; Wang, L.; Li, X.; Gong, H.; Shi, C.; Zhong, R.; Liu, X. Comparison of UAV and WorldView-2 imagery for mapping leaf area index of mangrove forest. *Int. J. Appl. Earth Obs. Geoinf.* **2017**, *61*, 22–31. [CrossRef]

98. Franklin, S.E.; Ahmed, O.S. Deciduous tree species classification using object-based analysis and machine learning with unmanned aerial vehicle multispectral data. *Int. J. Remote Sens.* **2017**, *39*, 5236–5245. [CrossRef]

99. Wallace, L.; Lucieer, A.; Malenovský, Z.; Turner, D.; Vopěnka, P. Assessment of forest structure using two UAV techniques: A comparison of airborne laser scanning and structure from motion (SfM) point clouds. *Forests* **2016**, *7*. [CrossRef]

100. Zhang, J.; Hu, J.; Lian, J.; Fan, Z.; Ouyang, X.; Ye, W. Seeing the forest from drones: Testing the potential of lightweight drones as a tool for long-term forest monitoring. *Biol. Conserv.* **2016**, *198*, 60–69. [CrossRef]

101. Guo, Q.; Su, Y.; Hu, T.; Zhao, X.; Wu, F.; Li, Y.; Liu, J.; Chen, L.; Xu, G.; Lin, G.; et al. An integrated UAV-borne lidar system for 3D habitat mapping in three forest ecosystems across China. *Int. J. Remote Sens.* **2017**, *38*, 2954–2972. [CrossRef]

102. Messinger, M.; Asner, G.P.; Silman, M. Rapid assessments of amazon forest structure and biomass using small unmanned aerial systems. *Remote Sens.* **2016**, *8*. [CrossRef]

103. Bedell, E.; Leslie, M.; Fankhauser, K.; Burnett, J.; Wing, M.G.; Thomas, E.A. Unmanned aerial vehicle-based structure from motion biomass inventory estimates. *J. Appl. Remote Sens.* **2017**, *11*, 026026. [CrossRef]

104. Rödig, E.; Cuntz, M.; Heinke, J.; Rammig, A.; Huth, A. Spatial heterogeneity of biomass and forest structure of the Amazon rainforest: linking remote sensing, forest modeling and field inventory. *Glob. Ecol. Biogeogr.* **2017**. [CrossRef]

105. Sankey, T.; Donager, J.; McVay, J.; Sankey, J.B. UAV lidar and hyperspectral fusion for forest monitoring in the southwestern USA. *Remote Sens. Environ.* **2017**, *195*, 30–43. [CrossRef]

106. Wilson, A.M.; Silander, J.A.; Gelfand, A.; Glenn, J.H. Scaling up: Linking field data and remote sensing with a hierarchical model. *Int. J. Geogr. Inf. Sci.* **2011**, *25*, 509–521. [CrossRef]

107. Chen, J.; Yi, S.; Qin, Y.; Wang, X. Improving estimates of fractional vegetation cover based on UAV in alpine grassland on the Qinghai–Tibetan Plateau. *Int. J. Remote Sens.* **2016**, *37*, 1922–1936. [CrossRef]

108. Rivas-Torres, G.F.; Benítez, F.L.; Rueda, D.; Sevilla, C.; Mena, C.F. A methodology for mapping native and invasive vegetation coverage in archipelagos: An example from the Galápagos Islands. *Prog. Phys. Geogr.* **2018**, *42*, 83–111. [CrossRef]

109. Szantoi, Z.; Smith, S.E.; Strona, G.; Koh, L.P.; Wich, S.A. Mapping orangutan habitat and agricultural areas using Landsat OLI imagery augmented with unmanned aircraft system aerial photography. *Int. J. Remote Sens.* **2017**, *38*, 1–15. [CrossRef]

110. Laliberte, A.S.; Goforth, M.A.; Steele, C.M.; Rango, A. Multispectral remote sensing from unmanned aircraft: Image processing workflows and applications for rangeland environments. *Remote Sens.* **2011**, *3*, 2529–2551. [CrossRef]

111. Iqbal, F.; Lucieer, A.; Barry, K. Simplified radiometric calibration for UAS-mounted multispectral sensor. *Eur. J. Remote Sens.* **2018**. [CrossRef]

112. Paneque-Gálvez, J.; McCall, M.K.; Napoletano, B.M.; Wich, S.A.; Koh, L.P. Small drones for community-based forest monitoring: An assessment of their feasibility and potential in tropical areas. *Forests* **2014**, *5*, 1481–1507. [CrossRef]

113. Mlambo, R.; Woodhouse, I.H.; Gerard, F.; Anderson, K. Structure from motion (SfM) photogrammetry with drone data: A low cost method for monitoring greenhouse gas emissions from forests in developing countries. *Forests* **2017**, *8*. [CrossRef]

114. Hill, D.J.; Tarasoff, C.; Whitworth, G.E.; Baron, J.; Bradshaw, J.L.; Church, J.S. Utility of unmanned aerial vehicles for mapping invasive plant species: a case study on yellow flag iris (Iris pseudacorus L.). *Int. J. Remote Sens.* **2017**, *38*, 2083–2105. [CrossRef]

115. Müllerová, J.; Brůna, J.; Dvořák, P.; Bartaloš, T.; Vítková, M. Does the data resolution/origin matter? Satellite, airborne and UAV imagery to tackle plant invasions. *Int. Arch. Photogramm. Remote Sens. Spat. Inf. Sci.* **2016**, *41*, 903–908. [CrossRef]

116. Michez, A.; Piégay, H.; Jonathan, L.; Claessens, H.; Lejeune, P. Mapping of riparian invasive species with supervised classification of Unmanned Aerial System (UAS) imagery. *Int. J. Appl. Earth Obs. Geoinf.* **2016**, *44*, 88–94. [CrossRef]

117. Müllerová, J.; Bartaloš, T.; Brůna, J.; Dvořák, P.; Vítková, M. Unmanned aircraft in nature conservation: an example from plant invasions. *Int. J. Remote Sens.* **2017**, *38*, 2177–2198. [CrossRef]

118. Perroy, R.L.; Sullivan, T.; Stephenson, N. Assessing the impacts of canopy openness and flight parameters on detecting a sub-canopy tropical invasive plant using a small unmanned aerial system. *ISPRS J. Photogramm. Remote Sens.* **2017**, *125*, 174–183. [CrossRef]

119. Papakonstantinou, A.; Topouzelis, K.; Pavlogeorgatos, G. Coastline Zones Identification and 3D Coastal Mapping Using UAV Spatial Data. *ISPRS Int. J. Geo-Inf* **2016**. [CrossRef]

120. Gonçalves, J.A.; Henriques, R. UAV photogrammetry for topographic monitoring of coastal areas. *ISPRS J. Photogramm. Remote Sens.* **2015**. [CrossRef]

121. Ventura, D.; Bruno, M.; Jona Lasinio, G.; Belluscio, A.; Ardizzone, G. A low-cost drone based application for identifying and mapping of coastal fish nursery grounds. *Estuar. Coast. Shelf Sci.* **2016**, *171*. [CrossRef]

122. Casella, E.; Collin, A.; Harris, D.; Ferse, S.; Bejarano, S.; Parravicini, V.; Hench, J.L.; Rovere, A. Mapping coral reefs using consumer-grade drones and structure from motion photogrammetry techniques. *Coral Reefs* **2017**, *36*, 269–275. [CrossRef]

123. Ventura, D.; Bonifazi, A.; Gravina, M.F.; Belluscio, A.; Ardizzone, G. Mapping and Classification of Ecologically Sensitive Marine Habitats Using Unmanned Aerial Vehicle (UAV) Imagery and Object-Based Image Analysis (OBIA). *Remote Sens.* **2018**, *10*, 1331. [CrossRef]

124. Parsons, M.; Bratanov, D.; Gaston, K.J.; Gonzalez, F. UAVs, hyperspectral remote sensing, and machine learning revolutionizing reef monitoring. *Sensors* **2018**, *18*, 2026. [CrossRef]
125. Ballari, D.; Orellana, D.; Acosta, E.; Espinoza, A.; Morocho, V. Uav monitoring for enviromental management in galapagos islands. *Int. Arch. Photogramm. Remote Sens. Spat. Inf. Sci.* **2016**, *XLI-B1*, 1105–1111. [CrossRef]
126. Knoth, C.; Klein, B.; Prinz, T.; Kleinebecker, T. Unmanned aerial vehicles as innovative remote sensing platforms for high-resolution infrared imagery to support restoration monitoring in cut-over bogs. *Appl. Veg. Sci.* **2013**. [CrossRef]
127. Chabot, D.; Dillon, C.; Shemrock, A.; Weissflog, N.; Sager, E.P.S. Geo-Information An Object-Based Image Analysis Workflow for Monitoring Shallow-Water Aquatic Vegetation in Multispectral Drone Imagery. *ISPRS Int. J. Geo-Inf.* **2018**. [CrossRef]
128. Pande-Chhetri, R.; Abd-Elrahman, A.; Liu, T.; Morton, J.; Wilhelm, V.L. Object-based classification of wetland vegetation using very high-resolution unmanned air system imagery. *Eur. J. Remote Sens. ISSNOnline) J. Eur. J. Remote Sens.* **2017**, *50*, 2279–7254. [CrossRef]
129. Marcaccio, J.V.; Markle, C.E.; Chow-Fraser, P. Unmanned aerial vehicles produce high-resolution, seasonally-relevant imagery for classifying wetland vegetation. *Int. Arch. Photogramm. Remote Sens. Spat. Inf. Sci.* **2015**, *40*, 249–256. [CrossRef]
130. Boon, M.A.; Greenfield, R.; Tesfamichael, S. Wetland Assessment Using Unmanned Aerial Vehicle (Uav) Photogrammetry. *Int. Arch. Photogramm. Remote Sens. Spat. Inf. Sci.* **2016**, *XLI-B1*, 781–788. [CrossRef]
131. Capolupo, A.; Kooistra, L.; Berendonk, C.; Boccia, L.; Suomalainen, J. Estimating Plant Traits of Grasslands from UAV-Acquired Hyperspectral Images: A Comparison of Statistical Approaches. *ISPRS Int. J. Geo-Inf.* **2015**. [CrossRef]
132. Lu, B.; He, Y. Species classification using Unmanned Aerial Vehicle (UAV)-acquired high spatial resolution imagery in a heterogeneous grassland. *ISPRS J. Photogramm. Remote Sens.* **2017**, *128*, 73–85. [CrossRef]
133. Mayr, M.J.; Malß, S.; Ofner, E.; Samimi, C. Disturbance feedbacks on the height of woody vegetation in a savannah: a multi-plot assessment using an unmanned aerial vehicle (UAV). *Int. J. Remote Sens.* **2017**, *39*, 4761–4785. [CrossRef]
134. Cruzan, M.B.; Weinstein, B.G.; Grasty, M.R.; Kohrn, B.F.; Hendrickson, E.C.; Arredondo, T.M.; Thompson, P.G. Small Unmanned Aerial Vehicles (Micro-UAVs, Drones) in Plant Ecology. *Appl. Plant Sci.* **2016**, *4*, 1600041. [CrossRef]
135. Kraaijenbrink, P.D.A.; Shea, J.M.; Pellicciotti, F.; de Jong, S.M.; Immerzeel, W.W. Object-based analysis of unmanned aerial vehicle imagery to map and characterise surface features on a debris-covered glacier. *Remote Sens. Environ.* **2016**, *186*, 581–595. [CrossRef]
136. Seier, G.; Kellerer-Pirklbauer, A.; Wecht, M.; Hirschmann, S.; Kaufmann, V.; Lieb, G.K.; Sulzer, W. UAS-based change detection of the glacial and proglacial transition zone at Pasterze Glacier, Austria. *Remote Sens.* **2017**. [CrossRef]
137. Burns, P.; Nolin, A. Using atmospherically-corrected Landsat imagery to measure glacier area change in the Cordillera Blanca, Peru from 1987 to 2010. *Remote Sens. Environ.* **2014**. [CrossRef]
138. Fraser, R.H.; Olthof, I.; Lantz, T.C.; Schmitt, C. UAV photogrammetry for mapping vegetation in the low-Arctic. *Arct. Sci.* **2016**, *2*, 79–102. [CrossRef]
139. Malenovský, Z.; Lucieer, A.; King, D.H.; Turnbull, J.D.; Robinson, S.A. Unmanned aircraft system advances health mapping of fragile polar vegetation. *Methods Ecol. Evol.* **2017**. [CrossRef]
140. Birdsong, T.W.; Bean, M.; Grabowski, T.B.; Hardy, T.B.; Heard, T.; Holdstock, D.; Kollaus, K.; Magnelia, S.; Tolman, K. Application and utility of a low-cost unmanned aerial ayistem to manage and conserve aquatic resources in four Texas rivers. *J. Southeast. Assoc. Fish Wildl. Agencies* **2015**, *2*, 80–85.
141. van Iersel, W.; Straatsma, M.; Middelkoop, H.; Addink, E. Multitemporal Classification of River Floodplain Vegetation Using Time Series of UAV Images. *Remote Sens.* **2018**, *10*, 1144. [CrossRef]
142. Woodget, A.S.; Austrums, R.; Maddock, I.P.; Habit, E. Drones and digital photogrammetry: from classifications to continuums for monitoring river habitat and hydromorphology. *Wiley Interdiscip. Rev. Water* **2017**, *4*, e1222. [CrossRef]
143. Husson, E.; Hagner, O.; Ecke, F. Unmanned aircraft systems help to map aquatic vegetation. *Appl. Veg. Sci.* **2013**, *17*, 567–577. [CrossRef]
144. Hilborn, R.; Arcese, P.; Borner, M.; Hando, J.; Hopcraft, G.; Loibooki, M.; Mduma, S.; Sinclair, A.R.E. Effective Enforcement in a Conservation Area. *Science* **2006**, *314*, 1266. [CrossRef]

145. Edgar, G.J.; Stuart-Smith, R.D.; Willis, T.J.; Kininmonth, S.; Baker, S.C.; Banks, S.; Barrett, N.S.; Becerro, M.; Bernard, A.T.F.; Berkhout, J.; Buxton, C.D.; et al. Global conservation outcomes depend on marine protected areas with five key features. *Nature* **2014**, *506*, 216. [CrossRef]

146. Struhsaker, T.T.; Struhsaker, P.J.; Siex, K.S. Conserving Africa's rain forests: Problems in protected areas and possible solutions. *Biol. Conserv.* **2005**. [CrossRef]

147. Di Franco, A.; Thiriet, P.; Di Carlo, G.; Dimitriadis, C.; Francour, P.; Gutiérrez, N.L.; De Grissac, A.J.; Koutsoubas, D.; Milazzo, M.; Otero, M.; Piante, C.; Plass-johnson, J.; Sainz-trapaga, S.; et al. Five key attributes can increase marine protected areas performance for small-scale fisheries management. *Sci. Rep.* **2016**, 1–9. [CrossRef] [PubMed]

148. Olivares-Mendez, M.A.; Bissyandé, T.F.; Somasundar, K.; Klein, J.; Voos, H.; Le Traon, Y. The NOAH Project: Giving a Chance to Threatened Species in Africa with UAVs. In *e-Infrastructure and e-Services for Developing Countries. AFRICOMM 2013*; Bissyandé, T., van Stam, G., Eds.; Lecture Notes of the Institute for Computer Sciences, Social Informatics and Telecommunications Engineering; Springer: Cham, Switzerland, 2014; Volume 135, pp. 198–208. [CrossRef]

149. Shaffer, M.J.; Bishop, J.A. Predicting and Preventing Elephant Poaching Incidents through Statistical Analysis, GIS-Based Risk Analysis, and Aerial Surveillance Flight Path Modeling. *Trop. Conserv. Sci.* **2016**, *9*, 525–548. [CrossRef]

150. Sabella, G.; Viglianisi, F.M.; Rotondi, S.; Brogna, F. Preliminary observations on the use of drones in the environmental monitoring and in the management of protected areas. The case study of "R.N.O. Vendicari", Syracuse (Italy). *Biodiversity J.* **2017**, *8*, 79–86.

151. Weber, S.; Knaus, F. Using drones as a monitoring tool to detect evidence of winter sports activities in a protected mountain area. *Eco.mont* **2017**, *9*, 30–34. [CrossRef]

152. Bondi, E.; Fang, F.; Hamilton, M.; Kar, D.; Dmello, D.; Choi, J.; Hannaford, R.; Iyer, A.; Joppa, L.; Tambe, M. SPOT poachers in action: Augmenting conservation drones with automatic detection in near real time. In Proceedings of the AAAI Conference on Artificial Intelligence/Thirty-Second AAAI Conference on Artificial Intelligence, New Orleans, LA, USA, 2–7 February 2018.

153. Kamminga, J.; Ayele, E.; Meratnia, N.; Havinga, P. Poaching detection technologies-A survey. *Sensors* **2018**, *18*. [CrossRef] [PubMed]

154. Banzi, J.F. A Sensor Based Anti-Poaching System in Tanzania. *Int. J. Sci. Res. Publ.* **2014**, *4*, 1–7.

155. Finn, R.L.; Wright, D. Privacy, data protection and ethics for civil drone practice: A survey of industry, regulators and civil society organisations. *Comput. Law Secur. Rev.* **2016**, *32*, 577–586. [CrossRef]

156. Duffy, R. Waging a war to save biodiversity: The rise of militarized conservation. *Int. Aff.* **2014**, *90*, 819–834. [CrossRef]

157. Shrestha, Y.; Lapeyre, R. Modern Wildlife Monitoring Technologies: Conservationists versus Communities? A Case Study: The Terai-Arc Landscape, Nepal. *Conserv. Soc.* **2018**, *16*, 91–101. [CrossRef]

158. Tulloch, V.J.D.; Tulloch, A.I.T.; Visconti, P.; Halpern, B.S.; Watson, J.E.M.; Evans, M.C.; Auerbach, N.A.; Barnes, M.; Beger, M.; Chadès, I.; et al. Why do We map threats? Linking threat mapping with actions to make better conservation decisions. *Front. Ecol. Environ.* **2015**. [CrossRef]

159. Arefin, A.M.E. Proposal of a marine protected area surveillance system against illegal vessels using image sensing and image processing. *Acta Ecol. Sin.* **2018**. [CrossRef]

160. Toonen, H.M.; Bush, S.R. The digital frontiers of fisheries governance: fish attraction devices, drones and satellites. *J. Environ. Policy Plan.* **2018**, *0*, 1–13. [CrossRef]

161. Radjawali, I.; Pye, O.; Flitner, M. Recognition through reconnaissance? Using drones for counter-mapping in Indonesia. *J. Peasant Stud.* **2017**, *44*, 753–769. [CrossRef]

162. Radjawali, I.; Pye, O. Drones for justice: Inclusive technology and river-related action research along the Kapuas. *Geogr. Helv.* **2017**, *72*, 17–27. [CrossRef]

163. Samia, D.S.M.; Angeloni, L.M.; Bearzi, M.; Bessa, E.; Crooks, K.R.; D'Amico, M.; Ellenberg, U.; Geffroy, B.; Larson, C.L.; Loyola, R.; et al. Best Practices Toward Sustainable Ecotourism. In *Ecotourism's Promise and Peril*; Blumstein, D., Geffroy, B., Samia, D., Bessa, E., Eds.; Springer: Cham, Switzerland, 2017; pp. 153–178. [CrossRef]

164. King, L.M. Will drones revolutionise ecotourism? *J. Ecotourism* **2014**, *13*, 85–92. [CrossRef]

165. Chamata, J.E.; King, L.M. The Commercial Use of Drones in U.S. National Parks. *Int. Technol. Manag. Rev.* **2017**, *6*, 158–164. [CrossRef]

166. Themistocleous, K.; Ioannides, M.; Agapiou, A.; Hadjimitsis, D.G. The methodology of documenting cultural heritage sites using photogrammetry, UAV, and 3D printing techniques: the case study of Asinou Church in Cyprus. *Proc. SPIE* **2015**. [CrossRef]

167. Hansen, A.S. Applying visitor monitoring methods in coastal and marine areas – some learnings and critical reflections from Sweden. *Scand. J. Hosp. Tour.* **2016**, *2250*, 1–18. [CrossRef]

168. Park, K.; Ewing, R. The usability of unmanned aerial vehicles (UAVs) for measuring park-based physical activity. *Landsc. Urban Plan.* **2017**, *167*, 157–164. [CrossRef]

169. Leary, D. Drones on ice: an assessment of the legal implications of the use of unmanned aerial vehicles in scientific research and by the tourist industry in Antarctica. *Polar Rec.* **2017**. [CrossRef]

170. Office of Environment and Heritage. New South Wales National Parks and Wildlife Service (NPWS) Drones in parks policy. Available online: https://www.environment.nsw.gov.au/topics/parks-reserves-and-protected-areas/park-policies/drones-in-parks (accessed on 19 October 2017).

171. De Peyer, R. Drones are banned from Royal Parks amid "fears over impact on wildlife and visitor safety." LondonStandard. *LondonStandard*, 9 March 2015.

172. Su, T.C. Multispectral sensors carried on unmanned aerial vehicle (UAV) for trophic state mapping of the small reservoir in Kinmen, Taiwan. In Proceedings of the International Geoscience and Remote Sensing Symposium (IGARSS), Milan, Italy, 26–31 July 2015; pp. 5348–5351.

173. Ore, J.P.; Elbaum, S.; Burgin, A.; Detweiler, C. Autonomous aerial water sampling. *J. Field Robot.* **2015**, *32*, 1095–1113. [CrossRef]

174. Koparan, C.; Koc, A.B.; Privette, C.V.; Sawyer, C.B.; Sharp, J.L. Evaluation of a UAV-assisted autonomous water sampling. *Water* **2018**, *10*. [CrossRef]

175. Tóth, V.R. Monitoring Spatial Variability and Temporal Dynamics of Phragmites Using Unmanned Aerial Vehicles. *Front. Plant Sci.* **2018**, *9*, 1–11. [CrossRef] [PubMed]

176. Koparan, C.; Koc, A.B.; Privette, C.V.; Sawyer, C.B. In Situ Water Quality Measurements Using an Unmanned Aerial Vehicle (UAV) System. *Water* **2018**, *10*, 264. [CrossRef]

177. D'Oleire-Oltmanns, S.; Marzolff, I.; Peter, K.D.; Ries, J.B. Unmanned aerial vehicle (UAV) for monitoring soil erosion in Morocco. *Remote Sens.* **2012**, *4*, 3390–3416. [CrossRef]

178. Casella, E.; Rovere, A.; Pedroncini, A.; Stark, C.P.; Casella, M.; Ferrari, M.; Firpo, M. Drones as tools for monitoring beach topography changes in the Ligurian Sea (NW Mediterranean). *Geo-Marine Lett.* **2016**, *36*, 151–163. [CrossRef]

179. Casella, E.; Rovere, A.; Pedroncini, A.; Mucerino, L.; Casella, M.; Cusati, L.A.; Vacchi, M.; Ferrari, M.; Firpo, M. Study of wave runup using numerical models and low-altitude aerial photogrammetry: A tool for coastal management. *Estuar. Coast. Shelf Sci.* **2014**, *149*, 160–167. [CrossRef]

180. Mokroš, M.; Výbošťok, J.; Merganič, J.; Hollaus, M.; Barton, I.; Koreň, M.; Tomaštík, J.; Čerňava, J. Early stage forest windthrow estimation based on unmanned aircraft system imagery. *Forests* **2017**, *8*. [CrossRef]

181. Liu, C.-C.; Chen, P.-L.; Matsuo, T.; Chen, C.-Y. Rapidly responding to landslides and debris flow events using a low-cost unmanned aerial vehicle. *J. Appl. Remote Sens.* **2015**. [CrossRef]

182. Jaukovic, I. Unmanned Aerial Vehicles: A new tool for landslide risk assessment. **2017**, 1–7.

183. Török, Á.; Barsi, Á.; Bögöly, G.; Lovas, T.; Somogyi, Á.; Görög, P. Slope stability and rockfall assessment of volcanic tuffs using RPAS with 2-D FEM slope modelling. *Nat. Hazards Earth Syst. Sci.* **2018**. [CrossRef]

184. Izumida, A.; Uchiyama, S.; Sugai, T. Application of UAV-SfM photogrammetry and aerial LiDAR to a disastrous flood: multitemporal topographic measurement of a newly formed crevasse splay of the Kinu River, central Japan. *Nat. Hazards Earth Syst. Sci. Discuss.* **2016**, 1–22. [CrossRef]

185. U.S. Geological Survey Volcano Hazards Program. Available online: https://volcanoes.usgs.gov/vsc/movies/movie_174028.html (accessed on 27 September 2018).

186. Nakano, T.; Kamiya, I.; Tobita, M.; Iwahashi, J.; Nakajima, H. Landform monitoring in active volcano by UAV and SFM-MVS technique. *Int. Arch. Photogramm. Remote Sens. Spat. Inf. Sci.* **2014**, *XL-8*, 71–75. [CrossRef]

187. Messinger, M.; Silman, M. Unmanned aerial vehicles for the assessment and monitoring of environmental contamination: An example from coal ash spills. *Environ. Pollut.* **2016**, *218*, 889–894. [CrossRef]

188. Wing, M.G.; Burnett, J.D.; Sessions, J. Remote Sensing and Unmanned Aerial System Technology for Monitoring and Quantifying Forest Fire Impacts. *Int. J. Remote Sens. Appl.* **2014**, *4*, 18. [CrossRef]

189. Cruz, H.; Eckert, M.; Meneses, J.; Martínez, J.F. Efficient forest fire detection index for application in Unmanned Aerial Systems (UASs). *Sensors* **2016**, *16*. [CrossRef]

190. Lorah, P.; Ready, A.; Rinn, E. Using Drones to Generate New Data for Conservation Insights. *Int. J. Geospatial Environ. Res.* **2018**, *5*, 2.

191. Karaca, Y.; Cicek, M.; Tatli, O.; Sahin, A.; Pasli, S.; Beser, M.F.; Turedi, S. The potential use of unmanned aircraft systems (drones) in mountain search and rescue operations. *Am. J. Emerg. Med.* **2017**. [CrossRef]

192. Van Tilburg, C.; Brown, S.T.; Ferguson, M. First Report of Using Portable Unmanned Aircraft Systems (Drones) for Search and Rescue. *Wilderness Environ. Med.* **2017**, *15*, 12. [CrossRef]

193. Hengstmann, E.; Gräwe, D.; Tamminga, M.; Fischer, E.K. Marine litter abundance and distribution on beaches on the Isle of Rügen considering the influence of exposition, morphology and recreational activities. *Mar. Pollut. Bull.* **2017**, *115*, 297–306. [CrossRef] [PubMed]

194. Martin, C.; Parkes, S.; Zhang, Q.; Zhang, X.; McCabe, M.F.; Duarte, C.M. Use of unmanned aerial vehicles for efficient beach litter monitoring. *Mar. Pollut. Bull.* **2018**, *131*, 662–673. [CrossRef] [PubMed]

195. Deidun, A.; Gauci, A.; Lagorio, S.; Galgani, F. Optimising beached litter monitoring protocols through aerial imagery. *Mar. Pollut. Bull.* **2018**, *131*, 212–217. [CrossRef] [PubMed]

196. Gómez, C.; Green, D.R. Small unmanned airborne systems to support oil and gas pipeline monitoring and mapping. *Arab. J. Geosci.* **2017**, *10*. [CrossRef]

197. Martínez-de Dios, J.; de San Bernabé, A.; Viguria, A.; Torres-González, A.; Ollero, A. Combining Unmanned Aerial Systems and Sensor Networks for Earth Observation. *Remote Sens.* **2017**, *9*, 336. [CrossRef]

198. Glenn, N.F.; States, U.; Artigas, F.; Environmental, M.; States, U.; Temperton, V.M. Timing Is Important: Unmanned Aircraft vs. Satellite Imagery in Plant Invasion Monitoring. *Front. Plant. Sci.* **2017**, *8*, 1–13. [CrossRef]

199. Rodriguez, R.; Jenkins, D.; Leary, J. Enhancing Invasive Species Control with Unmanned Aerial Systems and Herbicide Ballistic Technology. *2017 ASABE Annual International Meeting* **2017**. [CrossRef]

200. Finn, R.L.; Wright, D. Unmanned aircraft systems: Surveillance, ethics and privacy in civil applications. *Comput. Law Secur. Rev.* **2012**, *28*, 184–194. [CrossRef]

201. Stöcker, C.; Bennett, R.; Nex, F.; Gerke, M.; Zevenbergen, J. Review of the Current State of UAV Regulations. *Remote Sens.* **2017**, *9*, 459. [CrossRef]

202. Cracknell, A.P. UAVs: regulations and law enforcement. *Int. J. Remote Sens.* **2017**, *38*, 3054–3067. [CrossRef]

203. Dormann, C.F.; McPherson, J.M.; Araújo, M.B.; Bivand, R.; Bolliger, J.; Carl, G.; Davies, R.G.; Hirzel, A.; Jetz, W.; Daniel Kissling, W.; et al. Methods to account for spatial autocorrelation in the analysis of species distributional data: A review. *Ecography* **2007**, *30*, 609–628. [CrossRef]

204. Wilson, R.P.; McMahon, C.R. Measuring devices on wild animals: What constitutes acceptable practice? *Front. Ecol. Environ.* **2006**, *4*, 147–154. [CrossRef]

205. Duriez, O.; Boguszewski, G.; Vas, E.; Gre, D. Approaching birds with drones: First experiments and ethical guidelines. *Biol. Lett.* **2015**. [CrossRef]

206. McEvoy, J.F.; Hall, G.P.; McDonald, P.G. Evaluation of unmanned aerial vehicle shape, flight path and camera type for waterfowl surveys: disturbance effects and species recognition. *PeerJ* **2016**, *4*. [CrossRef] [PubMed]

207. Fletcher, S.B.B.A.T. Will drones reduce investigator disturbance to surface-nesting birds? *Mar. Ornithol.* **2017**, *45*, 89–94.

208. Scobie, C.A.; Hugenholtz, C.H. Wildlife monitoring with unmanned aerial vehicles: Quantifying distance to auditory detection. *Wildl. Soc. Bull.* **2016**, *40*, 781–785. [CrossRef]

209. Lyons, M.; Brandis, K.; Callaghan, C.; Mccann, J.; Mills, C.; Ryall, S.; Kingsford, R. Bird interactions with drones from individuals to large colonies. *bioRxiv* **2017**, 1–10. [CrossRef]

210. Bevan, E.; Whiting, S.; Tucker, T.; Guinea, M.; Raith, A.; Douglas, R. Measuring behavioral responses of sea turtles, saltwater crocodiles, and crested terns to drone disturbance to define ethical operating thresholds. *PLoS One* **2018**, *13*, 1–17. [CrossRef] [PubMed]

211. Ditmer, M.A.; Vincent, J.B.; Werden, L.K.; Tanner, J.C.; Laske, T.G.; Iaizzo, P.A.; Garshelis, D.L.; Fieberg, J.R. Bears Show a Physiological but Limited Behavioral Response to Unmanned Aerial Vehicles. *Curr. Biol.* **2015**, *25*, 2278–2283. [CrossRef] [PubMed]

212. Pomeroy, P.; O'Connor, L.; Davies, P. Assessing use of and reaction to unmanned aerial systems in gray and harbor seals during breeding and molt in the UK. *J. Unmanned Veh. Syst.* **2015**, *3*, 102–113. [CrossRef]

213. Ramos, E.A.; Maloney, B.M.; Magnasco, M.O.; Reiss, D. Bottlenose Dolphins and Antillean Manatees Respond to Small Multi-Rotor Unmanned Aerial Systems. *Front. Mar. Sci.* **2018**, *5*, 316. [CrossRef]

214. Hodgson, J.C.; Koh, L.P. Best practice for minimising unmanned aerial vehicle disturbance to wildlife in biological field research. *Curr. Biol.* **2016**, *26*. [CrossRef] [PubMed]

215. Gonzalez, F.; Johnson, S. Standard operating procedures for UAV or drone based monitoring of wildlife. Proceedings of UAS4RS 2017 (Unmanned Aircr. Syst. Remote Sensing), Hobart, TAS, Australia, 24–25 May 2017; pp. 1–8.

216. Grémillet, D.; Puech, W.; Garçon, V.; Boulinier, T.; Le Maho, Y. Robots in ecology: welcome to the machine. *Open J. Ecol.* **2012**, *2*, 49–57. [CrossRef]

217. Sepúlveda, A.; Schluep, M.; Renaud, F.G.; Streicher, M.; Kuehr, R.; Hagelüken, C.; Gerecke, A.C. A review of the environmental fate and effects of hazardous substances released from electrical and electronic equipments during recycling: Examples from China and India. *Environ. Impact Assess. Rev.* **2010**, *30*, 28–41. [CrossRef]

218. AUVSI Are UAS More Cost Effective than Manned Flights? | Association for Unmanned Vehicle Systems International. Available online: http://www.auvsi.org/are-uas-more-cost-effective-manned-flights (accessed on 15 October 2017).

219. Koski, W.; Abgrall, P.; Yazvenko, S. An Inventory and Evaluation of Unmanned Aerial Systems for Offshore Surveys of Marine Mammals. *J. Cetacean Res. Manag.* **2011**, *11*, 239–247.

220. Liao, X.; Zhang, Y.; Su, F.; Yue, H.; Ding, Z.; Liu, J. UAVs surpassing satellites and aircraft in remote sensing over China. *Int. J. Remote Sens.* **2018**, *00*, 1–16. [CrossRef]

221. Otero, V.; Van De Kerchove, R.; Satyanarayana, B.; Martínez-Espinosa, C.; Bin Fisol, M.A.; Bin Ibrahim, M.R.; Sulong, I.; Mohd-Lokman, H.; Lucas, R.; Dahdouh-Guebas, F. Managing mangrove forests from the sky: Forest inventory using field data and Unmanned Aerial Vehicle (UAV) imagery in the Matang Mangrove Forest Reserve, peninsular Malaysia. *For. Ecol. Manage.* **2018**, *411*, 35–45. [CrossRef]

222. Andrew, M.E.; Shephard, J.M. Semi-automated detection of eagle nests: an application of very high-resolution image data and advanced image analyses to wildlife surveys. *Remote Sens. Ecol. Conserv.* **2017**. [CrossRef]

223. Chabot, D.; Francis, C.M. Computer-automated bird detection and counts in high-resolution aerial images: a review. *J. F. Ornithol.* **2016**, *87*, 343–359. [CrossRef]

224. Gonzalez, L.F.; Montes, G.A.; Puig, E.; Johnson, S.; Mengersen, K.; Gaston, K.J. Unmanned aerial vehicles (UAVs) and artificial intelligence revolutionizing wildlife monitoring and conservation. *Sensors* **2016**, *16*. [CrossRef]

225. Lhoest, S.; Linchant, J.; Quevauvillers, S.; Vermeulen, C.; Lejeune, P. How many hippos (Homhip): Algorithm for automatic counts of animals with infra-red thermal imagery from UAV. *Int. Arch. Photogramm. Remote Sens. Spat. Inf. Sci.* **2015**, *40*, 355–362. [CrossRef]

226. Martin, J.; Edwards, H.H.; Burgess, M.A.; Percival, H.F.; Fagan, D.E.; Gardner, B.E.; Ortega-Ortiz, J.G.; Ifju, P.G.; Evers, B.S.; Rambo, T.J. Estimating distribution of hidden objects with drones: From tennis balls to manatees. *PLoS One* **2012**, *7*, 1–8. [CrossRef] [PubMed]

227. Abd-Elrahman, A.; Pearlstine, L.; Percival, F. Development of Pattern Recognition Algorithm for Automatic Bird detection from unmanned aerial vehicle imagery. *Surv. L. Inf. Sci.* **2005**, *65*, 37.

228. Longmore, S.N.; Collins, R.P.; Pfeifer, S.; Fox, S.E.; Mulero-Pazmany, M.; Bezombes, F.; Goodwin, A.; De Juan Ovelar, M.; Knapen, J.H.; Wich, S.A. Adapting astronomical source detection software to help detect animals in thermal images obtained by unmanned aerial systems. *Int. J. Remote Sens.* **2017**, *38*, 2623–2638. [CrossRef]

229. Hodgson, J.C.; Mott, R.; Baylis, S.M.; Pham, T.T.; Wotherspoon, S.; Kilpatrick, A.D.; Raja Segaran, R.; Reid, I.; Terauds, A.; Koh, L.P. Drones count wildlife more accurately and precisely than humans. *Methods Ecol. Evol.* **2018**, *9*, 1160–1167. [CrossRef]

230. Ma, L.; Li, M.; Ma, X.; Cheng, L.; Du, P.; Liu, Y. A review of supervised object-based land-cover image classification. *ISPRS J. Photogramm. Remote Sens.* **2017**, *130*, 277–293. [CrossRef]

231. Hill, S.L.; Clemens, P. Miniaturization of high spectral spatial resolution hyperspectral imagers on unmanned aerial systems. *Proc. SPIE* **2015**, *9482*. [CrossRef]

232. Lucieer, A.; Malenovsk??, Z.; Veness, T.; Wallace, L. HyperUAS - Imaging spectroscopy from a multirotor unmanned aircraft system. *J. F. Robot.* **2014**. [CrossRef]

233. Turner, D.; Lucieer, A.; Watson, C. An automated technique for generating georectified mosaics from ultra-high resolution Unmanned Aerial Vehicle (UAV) imagery, based on Structure from Motion (SFM) point clouds. *Remote Sens.* **2012**, *4*, 1392–1410. [CrossRef]

234. Duarte, L.; Teodoro, A.C.; Moutinho, O.; Gonçalves, J.A. Open-source GIS application for UAV photogrammetry based on MicMac. *Int. J. Remote Sens.* **2017**, *38*, 3181–3202. [CrossRef]

235. Gonçalves, G.R.; Pérez, J.A.; Duarte, J. Accuracy and effectiveness of low cost UASs and open source photogrammetric software for foredunes mapping. *Int. J. Remote Sens.* **2018**, *39*, 5059–5077. [CrossRef]

236. Baena, S.; Boyd, D.S.; Moat, J. UAVs in pursuit of plant conservation - Real world experiences. *Ecol. Inform.* **2017**, 2–9. [CrossRef]

237. Lee, J.G.; Kang, M. Geospatial Big Data: Challenges and Opportunities. *Big Data Res.* **2015**. [CrossRef]

238. Solpico, D.B.; Libatique, N.J.C.; Tangonan, G.L.; Cabacungan, P.M.; Girardot, G.; Ezequiel, C.A.F.; Favila, C.M.; Honrado, J.L.E.; Cua, M.A.; Perez, T.R.; et al. Towards a web-based decision system for Philippine lakes with UAV imaging, water quality wireless network sensing and stakeholder participation. In Proceedings of the 2015 IEEE 10th International Conference on Intelligent Sensors, Sensor Networks and Information Processing, ISSNIP 2015, Singapore, 7–9 April 2015.

239. Popescu, D.; Ichim, L.; Stoican, F. Unmanned Aerial Vehicle Systems for Remote Estimation of Flooded Areas Based on Complex Image Processing. *Sensors* **2017**, *17*, 446. [CrossRef] [PubMed]

240. Tay, J.Y.L.; Erfmeier, A.; Kalwij, J.M. Reaching new heights: can drones replace current methods to study plant population dynamics? *Plant Ecol.* **2018**, *8*. [CrossRef]

241. Berra, E.F.; Gaulton, R.; Barr, S. Commercial Off-The-Shelf Digital Cameras on Unmanned Aerial Vehicles for Multi- Temporal Monitoring of Vegetation Reflectance and NDVI. *IEEE Trans. Geosci. Remote Sens.* **2017**, 1–35.

242. Faye, E.; Rebaudo, F.; Yánez-Cajo, D.; Cauvy-Fraunié, S.; Dangles, O. A toolbox for studying thermal heterogeneity across spatial scales: From unmanned aerial vehicle imagery to landscape metrics. *Methods Ecol. Evol.* **2016**. [CrossRef]

243. Westoby, M.J.; Brasington, J.; Glasser, N.F.; Hambrey, M.J.; Reynolds, J.M. "Structure-from-Motion" photogrammetry: A low-cost, effective tool for geoscience applications. *Geomorphology* **2012**, *179*, 300–314. [CrossRef]

244. Dubois, G.; Clerici, M.; Jf, P.; Brink, A.; Palumbo, I.; Gross, D.; Peedell, S.; Simonetti, D.; Punga, M. On the contribution of remote sensing to DOPA, a digitial observatory for protected areas. In Proceedings of the Proceedings of the 34th International Symposium on Remote Sensing of Environment (ISRSE), Sydney, Australia, 10–15 April 2011.

245. Hart, J.K.; Martinez, K. Environmental Sensor Networks: A revolution in the earth system science? *Earth-Science Rev.* **2006**. [CrossRef]

246. Chape, S.; Harrison, J.; Spalding, M.; Lysenko, I. Measuring the extent and effectiveness of protected areas as an indicator for meeting global biodiversity targets. *Philos. Trans. R. Soc. B Biol. Sci.* **2005**, *360*, 443–455. [CrossRef]

![drones logo] *drones*

MDPI

Article

Enhancement of Ecological Field Experimental Research by Means of UAV Multispectral Sensing

Ricardo Díaz-Delgado [1,*], Gábor Ónodi [2], György Kröel-Dulay [2] and Miklós Kertész [2]

[1]   Remote Sensing and GIS Laboratory (LAST-EBD). Estación Biologica de Doñana. CSIC. Avda. Américo
      Vespucio 26, 41092 Sevilla, Spain
[2]   MTA Centre for Ecological Research, Institute of Ecology and Botany, Alkotmány 2-4,
      H-2163 Vácrátót, Hungary; onodi.gabor@okologia.mta.hu (G.O.);
      kroel-dulay.gyorgy@okologia.mta.hu (G.K.-D.); kertesz.miklos@okologia.mta.hu (M.K.)
*   Correspondence: rdiaz@ebd.csic.es; Tel.: +34-954-232-340

Received: 23 November 2018; Accepted: 28 December 2018; Published: 7 January 2019

Abstract: Although many climate research experiments are providing valuable data, long-term measurements are not always affordable. In the last decades, several facilities have secured long-term experiments, but few studies have incorporated spatial and scale effects. Most of them have been implemented in experimental agricultural fields but none for ecological studies. Scale effects can be assessed using remote sensing images from space or airborne platforms. Unmanned aerial vehicles (UAVs) are contributing to an increased spatial resolution, as well as becoming the intermediate scale between ground measurements and satellite/airborne image data. In this paper we assess the applicability of UAV-borne multispectral images to provide complementary experimental data collected at point scale (field sampling) in a long-term rain manipulation experiment located at the Kiskun Long-Term Socio-Ecological Research (LTSER) site named ExDRain to assess the effects on grassland vegetation. Two multispectral sensors were compared at different scales, the Parrot Sequoia camera on board a UAV and the portable Cropscan spectroradiometer. The NDVI values were used to assess the effect of plastic roofs and a proportional reduction effect was found for Sequoia-derived NDVI values. Acceptable and significant positive relationships were found between both sensors at different scales, being stronger at Cropscan measurement scale. Differences found at plot scale might be due to heterogeneous responses to treatments. Spatial variability analysis pointed out a more homogeneous response for plots submitted to severe and moderate drought. More investigation is needed to address the possible effect of species abundance on NDVI at plot scale contributing to a more consistent representation of ground measurements. The feasibility of carrying out systematic UAV flights coincident or close to ground campaigns will certainly reveal the consistency of the observed spatial patterns in the long run.

Keywords: unmanned aerial vehicles (UAVs); field experiments; LTSER; drought; multiscale approach; NDVI; Sequoia

---

1. Introduction

Climate change effects on ecosystems are being investigated at different spatial and temporal scales. Both observational and experimental approaches are being applied to identify trends, shifts and changes for different ecological indicators. Among the many initiatives, the Long-Term Ecological Research (LTER) networks are informing about the factors driving changes in biodiversity, the self-organizing capacity of ecosystems, the effects of rare events and disturbances, the impacts of stressors on ecosystem function and the interactions between short- and long-term trends [1]. These LTER networks rely on site-based monitoring and research by providing data and detecting trends identifying drivers and pressures on ecosystems and biota. The LTER networks' major contribution is the enlargement of temporal and spatial scales to test global research hypotheses. In this case, as in the NEON [2] and GCOS [3] networks, observational approach is the dominant method.

Experimental designs have been widely implemented to test research hypothesis in relation to climate change. Typically, field experiments are designed to replicate conditions at which single and multi-factor effects occur. Factors are controlled and effects are measured during a predetermined period or in few cases for the long run such as ANAEE [4] or Drought-Net [5]. At plot scale, climate manipulation experiments are a particularly effective way to study the ecological consequences of climate change [6]. However, they are normally conceived to collect measurements at a certain spatial scale, usually related to plot size or sampling method [4]. Only few experiments are planned which follow a multiscale approach. Historical examples are the Oregon Transect Ecosystem Research Project [7] or the HiWATER [8] which used remote sensing images together with ground-truth data. Nevertheless, the experimental ecologist is hard-pressed to find specific guidance for the design, execution and analysis of experiments to produce results that account for scale-dependent effects [9].

Remote sensing has traditionally been proposed as the essential tool to scale down the processes observed with the help of the images provided either by Earth observation satellites or airborne cameras/sensors and ground measurements. While covering the gradient from global to local scales, remote sensing has been providing critical information to map changes and trends [10]. In the last decades, remote sensing scientists are thoroughly investigating upscaling procedures to integrate multiscale information for any observation of ecological relevance [11,12]. Major advances have been reached in precision farming [13] mostly by using thermal, multi and hyperspectral airborne sensors. However, there is a lack of research in multiscale approaches using remote sensing for long-term ecological and climate change experiments [4]. There are many high and very high-resolution sensors on board of Earth observation satellites providing images with spatial resolution going from a few meters to tens of centimetres. Such availability enhances the multiscale approach but the high-resolution images are costly and have to be pre-ordered and acquired over the study area. High-, medium- and low-resolution scenes are periodically acquired by the orbiting satellites collecting a time series of images illustrating temporal changes and trends at the landscape scale. The fine resolution satellites or airborne campaigns may help in detailed habitat mapping, for instance while dramatically increasing costs. Yet, Unmanned Aerial Vehicles (UAVs) can be flown over the same area as frequently as required, only constrained by weather conditions or legislation, becoming a suitable tool to map either elements or processes. As a major trait, UAVs provide the opportunity to define spatial resolution as detailed as requested according to the mission objectives [14]. Remote sensing UAV is becoming very useful in cropland monitoring [15,16] and precision farming [17]. It is also being widely applied for environmental assessment [18] and similar studies on grasslands have addressed sensor comparison for leaf area index estimation with ground measurements and UAV acquisitions [19]. Few studies are focusing on upscaling essential variables from natural vegetation making use of UAV and multispectral sensors [14,20].

As an expected consequence of global warming, extreme events are becoming more frequent across ecosystems including extreme and sudden droughts causing vegetation die-offs and community shifts [21]. Extreme drought events can reduce primary production [22] and community functional diversity [23]. At plot scale, climate manipulation experiments are particularly effective way to study ecological consequences of climate change, especially long-term multi-site field experiments [6,24]. This is the case of the ExDRain experiment set out at Kiskun LTER station where rain manipulation is applied since 2014 for several plots after the application of extreme drought event for some of them [25]. Ecosystem recovery is assessed by periodically measuring plant cover and biomass using NDVI (Normalized Difference Vegetation Index, [26]) collected using a portable CropScan MSR87 multispectral radiometer (Cropscan, Inc., Rochester, MN, USA) as a non-destructive method [27]. Cropscan has 8 narrow (10 nm bandwidth) spectral bands centred at 460, 510, 560, 610, 660, 710, 760 and 810 nm (bands 660 and 810 are usually used for NDVI calculation). However, the plant community is composed by several grassland species and the response to treatments is very heterogeneous [6] due to different phenology and structure. Plots are larger than Cropscan footprint measurement and UAV complementary information might help in providing ancillary information enhancing the assessment on treatments effects.

In this work we assess the applicability of UAV-borne multispectral and RGB (Red-Green-Blue) cameras to enhance experimental data by providing complementary information. We check the correlation between data collected at point (field sampling) and plot (UAV sampling) scales and analyse new observed patterns in relation to the effects of the ExDRain experiment on the vegetation. We question whether NDVI-derived from UAV multispectral images acquired at lower scale are in coherence with Cropscan point measurements. On the other hand, we examine the information provided by UAV images and assess the effects of experimental treatments at plot scale, by which we hypothesize that effects at different scales may be different and can enhance the interpretation of the observed results.

2. Study Site and Experimental Design

2.1. Kiskun LTER Site

The Kiskun LTER site is located in the Kiskunság National Park (46°52′N, 19°25′E) in a Pannonian sand forest-steppe vegetation mosaic [25] of high plant diversity and nature protection value [28]. It is included into the Kiskun LTSER (Long-Term Socio-Ecological Research) platform (Figure 1). The Kiskunság region is located in Central Hungary covering 14,000 km². It is an extremely heterogeneous sandy area, consisting of arable fields, abandoned pastures, planted forests and extensive natural and semi-natural habitats from xeric grasslands to salt marshes. The soil is calcaric arenosol which enhances the semidesert character of the vegetation. Climate of the study area is temperate continental. The vegetation period starts in April and finishes in October. Based on regional 30 years average values (1961–1990), mean annual temperature is 10.4 °C, mean monthly temperature ranges from −1.9 °C in January to 21.1 °C in July, while mean annual precipitation is 505 mm with a peak in June [29].

Kiskun LTER focuses on studying the effects of climate change (more extreme weather events and longer vegetation periods) and land use change (abandonment of arable fields, decrease of grazing and afforestation) on biodiversity as well as ecosystem functions and services in Kiskunság. According to climate change scenarios for Hungary, the frequency of extreme dry and wet years is expected to increase in the study region [30].

2.2. ExDRain Experiment

The experimental area represents the sand grassland of the continental semiarid forest-steppe biome of Central Europe. The purpose of the experiment is to investigate how extreme and moderate events interact by observing their single and combined effects on plant cover, abundance and biomass of the grassland species in the site. The experimental design takes the results of the multi-site EU FP5 VULCAN and the EU FP7 INCREASE projects [24,31,32] into consideration. In our study plots, we sampled open grassland patches of semi-arid perennial grassland dominated by C3 bunchgrasses. We study the ecosystem recovery following the extreme drought and how it is affected by experimentally reduced or increased precipitation.

One-off extreme drought treatment was created by excluding precipitation for five months in 2014. Starting in 2015 and repeating in each summer, four levels of long-term precipitation change are applied: (1) strong drought (2-months duration), (2) moderate drought (1-month), (3) control, and (4) water addition (four times per year, one per month between May and August, ca. 100 mm in total). The two (i.e., extreme and precipitation change) treatments are combined in a full factorial design ($2 \times 4 = 8$ treatment combinations), in six replications (Figure 1c,d) resulting in a total number of 48 plots of 3×3 m size. Exclusion is reached by covering with plastic the corresponding plots and watering using irrigation diffusors (Figure 2).

Plant biomass was estimated by non-destructive field spectroscopy with a portable Cropscan MSR87 multispectral radiometer (Cropscan, Inc., Rochester, MN, USA) measuring at the same time incoming and reflected radiation. Cropscan measurements were taken always exactly at the same location placing the sensor above the plots at a height of 1.5 m. With a field of view (FOV) of 28°, the Cropscan samples a circular area of 0.44 m^2 (diameter: 0.75 m) and the distance between measurement points of the neighbouring plots was 1 m. The frame allowed us to repeat the sampling of each plot at the same position during the different measurement events. Plastic roofs were removed for every Cropscan measurement event. The last measurement for this study took place during the week of 23 to 27 July 2018.

From red (660 nm) and near infrared (810 nm) Cropscan reflectance values bands, we calculated NDVI. The NDVI provides an accurate proxy for plant aboveground green biomass estimation for pioneering plant communities [27,33].

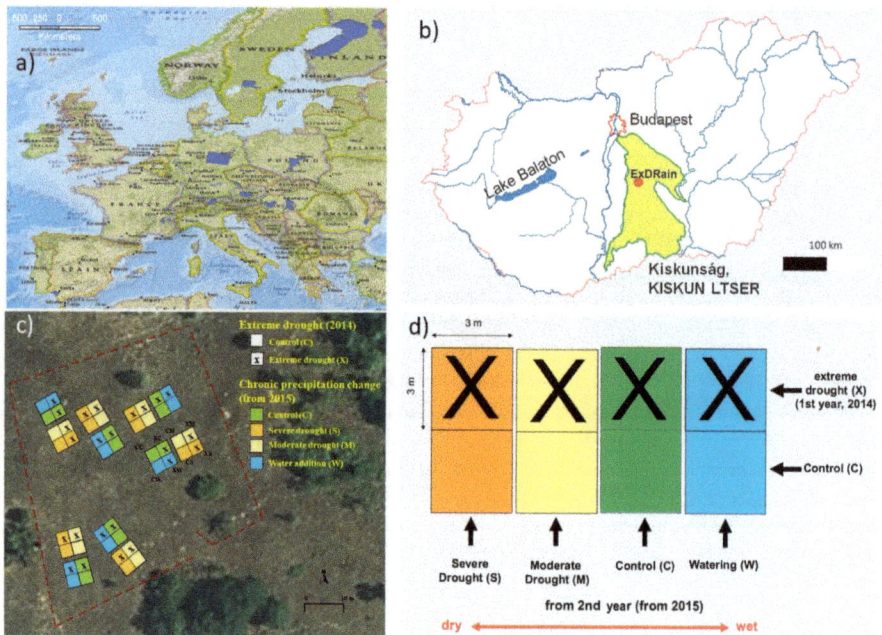

Figure 1. (**a**) Location of the 16 first Long-Term Socio-Ecological Research (LTSER) platforms declared in 2007 in Long-Term Ecological Research (LTER)-Europe [34] plus the three incorporated in 2010. Nowadays, the number has increased to 31 [35] and more have been created in other LTER regional networks [36]. (**b**) Location of Kiskun LTSER platform in Hungary (green area) and the ExDRain experiment site (red dot). (**c**) Details of experimental area with the different treatment plots. (**d**) Plot size and treatment legend by colours and crosses.

Figure 2. Left picture shows the removable plastic roofs used to exclude from rainfall for drought treatments in the ExDRain experiment and right picture the irrigating event for the watered plots.

3. Materials and Methods

3.1. UAV and Equipment

We used DJI Phantom 4+ quadcopter equipped with its original 4K 20 Mpix RGB CMOS camera plus a special mount designed to bring the camera Parrot Sequoia (Figure 3). This multispectral sensor captures images at 4 spectral bands (b1 green −550@40 nm, b2 red −660@40 nm, b3 red edge −735@10 nm and b4 near infrared −790@40 nm) with a vertical FOV of 48.5°, in addition to an RGB sensor which was shut off during this study. The Sequoia camera is connected to its own battery and provides wireless connection to be accessed and programmed through a computer. Additionally, the Sequoia camera brings a sensor of irradiance located in the upper part of the mount which is concurrently capturing irradiance while taking pictures [37]. A calibration panel is provided with every Sequoia camera to be pictured before flight allowing for bands' reflectance calculation after flight [38].

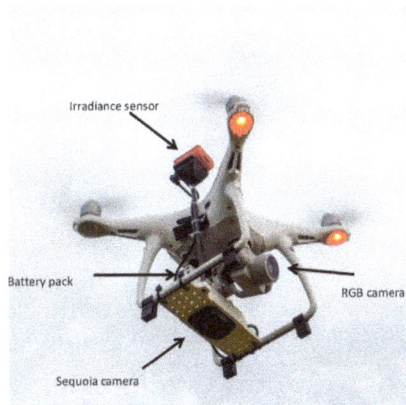

Figure 3. Unmanned Aerial Vehicle used for the study consisting of a DJI Phantom 4+ quadcopter equipped with its 4K RGB camera and a specific mount for the Parrot Sequoia multispectral camera, including both the camera itself and the sunshine sensor (original design from Zcopters).

3.2. Mission Planning and Geometric Processing

On 30 July at solar noon and with clear sky conditions we flew the UAV Phantom 4 Pro + equipped with the Parrot Sequoia multispectral camera over the whole ExDRain area (0.31 ha from 3.86 ha total flight area, Figure 4). One single flight of 14 minutes was carried out at 50 m above ground seeking to produce output images of 5.50 cm of nominal ground sampling distance (GSD) for Sequoia bands and 1.43 for the RGB 4K camera. The flight was carried out at the lowest speed to increase platform stability [39,40]. Radiometric calibration was simply achieved by reflectance calculation according to radiance coefficients and irradiance measured at every picture centre [41]. Vignetting correction was applied in the process. Mission planning was carried out using Pix4DMapper© software for Android devices which allows for the design of grid missions and defining all flight parameters. Pictures lateral and longitudinal overlap was defined as 80% both for DJI Phantom 4 Pro+ mission and for Sequoia camera which calculates the corresponding time and distance between camera shots.

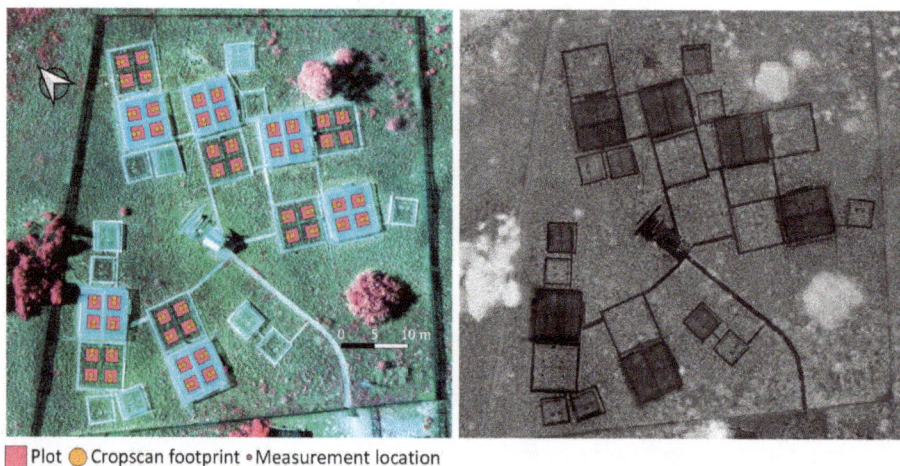

■ Plot ◯ Cropscan footprint ● Measurement location

Figure 4. Left image shows a false color RGB composite with near infrared-red edge-red Sequoia bands with the overlay of plot limits, Cropscan sampling footprints and measurement location points. Right image shows Normalized Difference Vegetation Index (NDVI) image calculated with Sequoia near infrared and red bands evidencing the plots covered by plastic (darker plots).

We located 16 ground control points (GCPs) with a differential GPS providing centimetric precision (20 cm on-site precision and 4 cm after post-processing) to improve geometric accuracy of the outputs to be produced. Images are automatically geotagged by Sequoia camera and the set of pictures were introduced into Pix4DMapper© software (Pix4D S.A., Lausanne, Switzerland) to be stitched and generate a multispectral orthomosaic together with digital surface model [42]. Ground control points are plotted on top of a point cloud at the intermediate step in the processing to provide a 3D Root Mean Square (RMS) error for the output. The NDVI image from Sequoia was generated using near infraredand red bands (Figure 4). Additionally, we processed the images acquired by the 4K RGB camera with a GSD of 1.43 cm to improve the geolocation of the periodical Cropscan point measurements.

3.3. Ground-Truth Sampling and Assessment of Plastic Effect

Cropscan periodical measurements were carried out during the previous week to the UAV flight when plastic roofs were removed. As data capture campaigns are too intensive and Cropscan measurements have to be also acquired close to noon, we could not fly with UAV and Sequoia camera on board exactly at the same time of acquisition. The experiment had to be strictly applied so that plastic roofs were mounted during UAV flight.

Therefore, as half of the plots were covered with plastic to exclude rain, we collected spectral signature of a calibration panel with Cropscan and Sequoia both above and below the plastic to assess plastic's effect on reflectance and irradiance. Image processing and data download was started once back at the office of the station.

Parallel ground measurements during UAV flight were also collected for several plots both with CropScan spectroradiometer and Sequoia camera in order to compare both sensors for the similar spectral wavelengths. These measurements were geolocated with sub-metric precision as well.

3.4. Multiscale Analysis and Spatial Variability

We compared NDVI Cropscan ground measurements for every plot with NDVI values from the Sequoia multispectral orthomosaic at 3 scales:

1. Pixel value at the point scale (n = 96).
2. Average value at the FOV scale (0.75 m diameter buffer around point measurements, n = 96).
3. Average value at the plot scale (3 × 3 m, n = 48).

The purpose of this analysis was to find the best related scale between both sensors and to assess sensor comparison. We used the lineal coefficient of determination R^2 for every scale comparison.

Finally, we compared spatial NDVI variability using standard deviation, Moran and Shannon Indices at plot scales taking advantage from the additional spatial information provided by Sequoia multispectral orthomosaic. This analysis will enable us to assess plot spatial heterogeneity, autocorrelation and how the Cropscan NDVI FOV measurements are related to Sequoia NDVI for plots identified per treatment.

4. Results

4.1. Geometric Accuracy of UAV Multispectral and RGB Orthomosaics

Table 1 shows the geometric characteristics of the two produced orthomosaics. Absolute root mean square errors (RMSEs) of the multispectral orthomosaic was below 1 pixel. However, it was bigger than 1 pixel for the RGB orthomosaic. Figure 4 shows a Sequoia false colour composite (NIR-RedEdge-Red) of the ExDRain experimental area and the NDVI image.

Table 1. Geometric characteristics of the UAV missions carried out over Doñana and Braila.

Flight Characteristics	Multispectral	RGB 4K Camera
Ground Sampling Distance (cm)	5. 5	1.45
Number of images	1064	135
Absolute RMS error (cm)	4.8	2.5

Location of Cropscan point measurements were revised and re-located using reference elements which were very conspicuous in both orthomosaics.

4.2. Plastic Effect

Spectral reflectance and irradiance captured above and below plastic roofs show overall lower values while measured under plastic (Figure 5). The effect on the irradiance is higher than for reflectance being magnified for short wavelengths.

The NDVI calculated from these measurements are higher for plastic covered plots when Cropscan was used below the plastic (R^2 = 0.81, $p < 0.01$, n = 8). Nevertheless, the relationship was lineal and underestimation was proportional to the measured values under plastic.

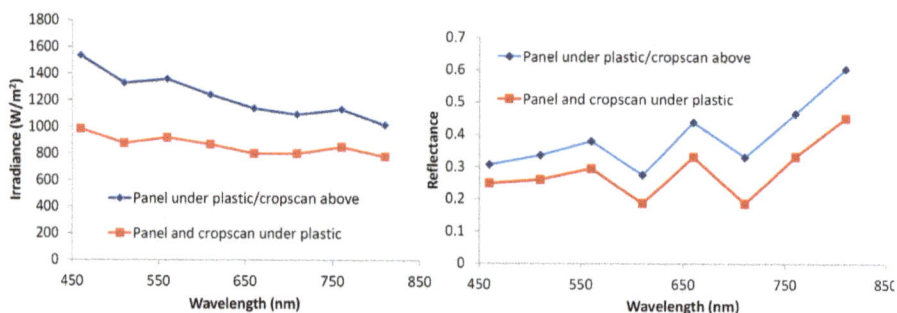

Figure 5. Left chart shows irradiance values collected with Cropscan spectroradiometer above and below plastic covered plots on a grey calibration panel. Right chart displays the corresponding reflectance values.

4.3. Multiscale Comparison

Table 2 shows the coefficient of determination (R^2) between the Cropscan and Sequoia NDVI values at different scales. Results are shown for all plots together and separately for plastic covered plots and uncovered ones. An acceptable positive and significant relationship was maintained between both sensors at the different scales, being lower at plot scale and higher at FOV scale for all measurements, covered and uncovered plots. Plastic covered plots systematically showed lower Sequoia NDVI values although proportional to those measured below plastic cover.

Table 2. R^2 values between Cropscan and Sequoia NDVI values for the different measurement scales ($n = 96$ for point and FOV measurements, $n = 48$ for plots). R^2 values between data from plastic covered and uncovered plots are also provided (sample size is divided by two). All R^2 values were significant at $p < 0.01$.

Scales	All Measurements	Plastic Cover	No Plastic
Point scale	0.43	0.31	0.37
FOV scale (circle, 40 cm radius)	0.46	0.41	0.65
Plot scale	0.38	0.21	0.33

Figure 6 shows the overall relationship between NDVI values of Cropscan and Sequoia at plot scale ($R^2 = 0.38$, $p < 0.01$, $n = 48$) with different markers as a function of the treatment.

We assessed treatments effect on Sequoia and Cropscan NDVI values. Aggregated Cropscan values (two FOV measurements per plot) were much more similar among treatments than Sequoia NDVI values at plot scale (Figure 7). Values for the covered plots were lower which can be due to the plastic effect but they still follow the measured trend by Cropscan, i.e., plots submitted to moderate drought show higher values than those under severe drought treatment. Yet, drought treatments significantly showed lower NDVI values than control and watered treatments (one sample *t*-test t = 2.79, $p < 0.01$, $n = 24$). Plots submitted to extreme drought effect did not show significant differences within treatment neither with Sequoia average NDVI values per plot nor with Cropscan point measurements (Figure 7).

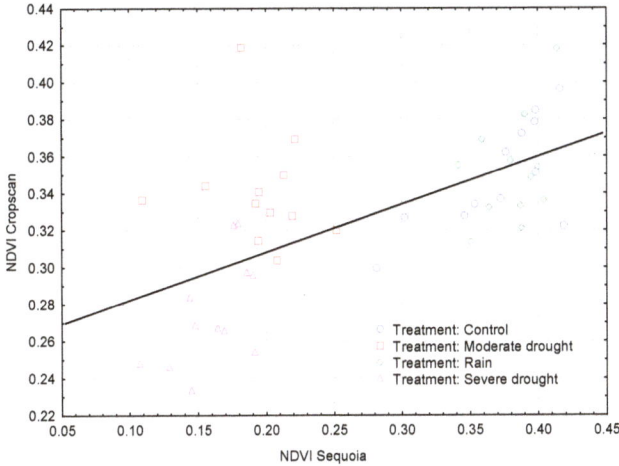

Figure 6. Scatterplot of Cropscan and Sequoia NDVI values for all plots. Different markers are used to distinguish among treatments. Linear fit is also shown ($R^2 = 0.38$, $p < 0.01$, $n = 48$).

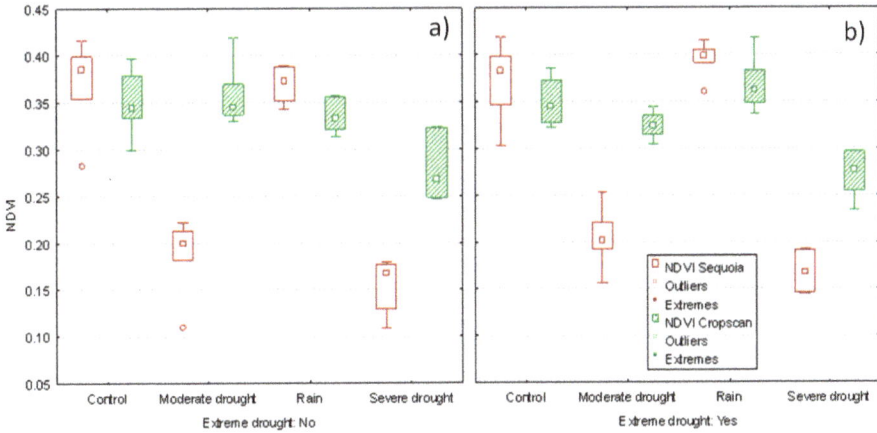

Figure 7. Boxplots of NDVI values from Cropscan and Sequoia plot measurements according to experimental treatments: control/moderate/severe drought/rainfed. Boxplot (**a**) shows the values for those plots that were not submitted to extreme drought in 2014 and boxplot (**b**) shows the values for those plots submitted to extreme drought.

We also assessed spatial variability within plots since Sequoia data provides much more information for every single plot than Cropscan. Figure 8 depicts the standard deviation for Sequoia NDVI values per plot indicating more spatial variability within control and watered plots than for plots submitted to moderate and severe drought.

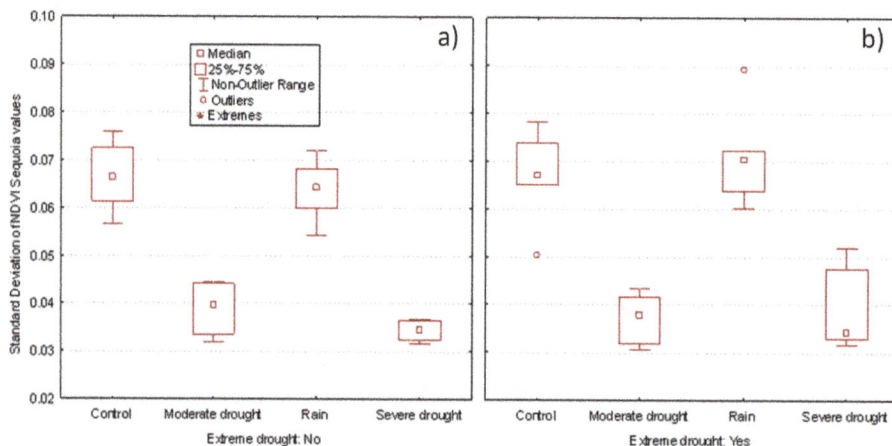

Figure 8. Boxplot of standard deviation for Sequoia NDVI values of ExDRain plots according to treatments. Boxplot (**a**) shows the values for those plots that were not submitted to extreme drought in 2014 and boxplot (**b**) shows the values for those plots submitted to extreme drought.

Average Moran Index values per treatment calculated from Sequoia NDVI values per plot revealed lower spatial autocorrelation for control and watered treatments than for both drought treatments (Table 3). Spatial autocorrelation of control and watered plots is lower, and therefore plots appear more heterogeneous in terms of NDVI variability. Conversely, either moderate or severe drought treatments manifest a higher homogeneity in NDVI values which might indicate a more homogeneous response to treatments. Yet, extreme drought event only is evidenced for severe drought treatment where the plots with the two treatments show higher spatial autocorrelation or a more homogeneous response. Shannon index did show a very low variability with a very similar pattern.

Table 3. Average Moran and Shannon Indices per treatment from NDVI Sequoia values at plot scale.

Treatment	Extreme Drought	Moran Index	Shannon Index
Control	No	0.8522	10.3499
	Yes	0.8473	10.3530
Moderate drought	No	0.9077	10.3525
	Yes	0.9043	10.3478
Rain	No	0.8440	10.3417
	Yes	0.8729	10.3522
Severe drought	No	0.8834	10.3462
	Yes	0.9195	10.3567

5. Discussion

Our study provides a first basis for the implementation of multiscale methodological approaches on field experimental sites. Rapid and easy deployment and flight of UAVs equipped with multispectral cameras can enhance and complement the results from these experiments by providing data from surrounding pixels not measured in the experiment. In our case, the use of Sequoia multispectral camera contributed to confirm the measured trends using Cropscan spectroradiometer at ground scale. The NDVI values retrieved from Sequoia measurements were significantly related to Cropscan measurements at different scales (Table 2), being weaker at plot scale and higher at FOV scale for uncovered plots and for all measurements. A lower relationship at plot scale may be revealing not necessarily an increase in sensor discrepancies but an influence of surrounding pixels in the treatment effect for the full plot. Such findings reveal a scale effect on treatments which has to be considered. The use of drone multispectral images clearly enhances the information of treatments effects on vegetation. So far, most of the multiscale approaches using UAVs have been applied in agricultural studies [18], showing that vegetation indices derived from the use of UAVs are reliable for assessing effects of experimental plots [43]. As UAVs are becoming widely available, its systematic use will certainly complement the data collected at ground scale. Much work has been done over crops evidencing the plausibility of upscaling essential information on crop health [15–17]. However, it is crucial to understand the critical difference between crops and natural grasslands, as essentially being manifested in the spatial distribution of individuals (homogeneous fields with similar growth and health versus heterogeneous growth, phenology and foliar greenness) and different species composition. Our main intention in this study was to show the spatial differences as a response to experimental treatments as an enhancement of such information by using UAVs.

One big constraint of the study was the presence of plastic roofs over the treated plots. Plastic effect on measurements consequently showed lower NDVI Sequoia values and higher Cropscan values measured under the plastic cover (due to reduced irradiance). Transparent plastic-mulch film allows visible lights to penetrate, but blocks outgoing long-wave radiation, and thus causes the greenhouse effect [44]. Levin et al. [45] claimed that white and transparent plastic-mulch films have three absorptions centred at 1218, 1732 and 2313 nm that are not affected by dust, rinse and surface factors. In our case, we did not use any of these wavelengths and plastic effect can be considered as negligible while making comparisons only among plastic covered plots.

Sequoia NDVI average values revealed much bigger differences between treatments than Cropscan FOV measurements (Figure 5), but this effect cannot be separated from the plastic effect as shown for Sequoia NDVI lower values. Significant differences are found between drought treatments (severe and moderate) and control/watered treatments for Sequoia. Extreme drought effect was evident neither with Cropscan nor with Sequoia measurements. These results are compliance with major findings from the VULCAN/INCREASE experiment which revealed that recurring drought treatment (precipitation exclusion) leads to mid-season reduction of aboveground biomass [6]. This is coherent with the results obtained by using NDVI from multispectral Sequoia on board of UAV.

Species composition has also been assessed under ExDRain experiment and it changed considerably in response to both extreme drought and long-term precipitation changes (unpublished data). Although UAV mapping could not deal with species mapping in this case, plant species variability might have an effect on NDVI values and therefore on spatial NDVI variability [46]. Actually, we were able to detect spatial variability within plots manifested for the treatments as more homogeneous response in plots under drought treatment according to SD and Moran Autocorrelation Index (Figure 8 and Table 3). Such finding may indicate heterogeneous responses to treatments based on the NDVI values at plot scale which are not considered at Cropscan FOV scale of measurement. Spatial variability might be due to different abundance on plant species and their spatial distribution patterns within plots, but this has to be investigated. Sha et al. [19] found better estimates of Leaf Area Index (LAI) values over sparsely vegetated areas as in our case. On the other hand, it also points out the need to assess the homogeneity of the measured effect per plot and per treatment in order to increase the consistency of the results. An analogous effect has been found for the temporal scale in the VULCAN/INCREASE experiment where results show that the effect of drought treatment can be overestimated with only one measurement at the time of the peak biomass, while multiple within-year measurements better describe the response of biomass [27].

One major challenge for these studies, where point measurements have to be located in high resolution orthomosaics, is the need to precisely geo-locate such measurements to guarantee a reliable data comparison. Even though high precision DGPS were used providing centimetric precision, RMS errors from the orthomosaics are close or are even bigger than final GSD. Therefore, the combined use of multispectral with RGB high-resolution images can clearly assist in re-locating the point measurements by identifying conspicuous elements in the image.

6. Conclusions

In this paper we confirmed the valuable and fast applicability of multispectral images captured by UAVs over experimental sites in providing complementary spatial information to the ground measurements at point scale. One single flight was used to confirm acceptable multiscale NDVI correlations being measured at point, FOV or plot scale. Although plastic covered plots reduced the NDVI values, main differences between treatments were coherent with the ones measured at ground scale with portable Cropscan spectroradiometer. Differences found at plot scale might be due to heterogeneous response to treatments. Spatial variability analysis pointed out a more homogeneous response for plots submitted to severe and moderate drought. More investigation is needed to address the possible effect on NDVI of species abundance at plot scale contributing to a more consistent representation of ground measurements. The feasibility to carry out systematic UAV flights in the future coincident or close to ground campaigns will certainly reveal the consistency of the observed spatial patterns in the long run.

Author Contributions: Conceptualization, R.D.-D. and M.K.; Data curation, R.D.-D., G.Ó. and M.K.; Formal analysis, R.D.-D.; Funding acquisition, R.D.-D. and M.K.; Investigation, R.D.-D., Gy.K.-D. and M.K.; Methodology, R.D.-D., G.Ó. and M.K.; Project administration, Gy.K.-D. and M.K.; Resources, G.Ó. and M.K.; Supervision, M.K.; Validation, R.D.-D., G.Ó. and M.K.; Visualization, G.Ó.; Writing—original draft, R.D.-D.; Writing—review & editing, R.D.-D..

Funding: This research was funded by the European Union's Horizon 2020 Research and Innovation Program under grant agreement No. 654359 (eLTER Horizon 2020 project). Gy. K-D. was supported by the National Research, Development and Innovation Fund (NRDI Fund) of Hungary (Nos. K112576, K129068).

Acknowledgments: The authors want to thank the funding by the European Union's Horizon 2020 Research and Innovation Program under grant agreement No. 654359 (eLTER Horizon 2020 project). Gy. K-D. was supported by the National Research, Development and Innovation Fund (NRDI Fund) of Hungary (Nos. K112576, K129068). We are grateful to the Hungarian Academy of Sciences which provided permits for fieldwork in the Kiskun ExDRain experiment.

Conflicts of Interest: The authors declare no conflict of interest. The founding sponsors had no role in the design of the study; in the collection, analyses, or interpretation of data; in the writing of the manuscript, and in the decision to publish the results.

References

1. Haase, P.; Tonkin, J.D.; Stoll, S.; Burkhard, B.; Frenzel, M.; Geijzendorffer, I.R.; Häuser, C.; Klotz, S.; Kühn, I.; McDowell, W.H.; et al. The next generation of site-based long-term ecological monitoring: Linking essential biodiversity variables and ecosystem integrity. *Sci. Total Environ.* **2018**, *613–614*, 1376–1384. [CrossRef] [PubMed]
2. Schimel, D.; Hargrove, W.; Hoffman, F.; MacMahon, J. NEON: A hierarchically designed national ecological network. *Front. Ecol. Environ.* **2007**, *5*, 59. [CrossRef]
3. Peterson, T.; Daan, H.; Jones, P. Initial Selection of a GCOS Surface Network. *Bull. Am. Meteorol. Soc.* **1997**, *78*, 2145–2152. [CrossRef]
4. Clobert, J.; Chanzy, A.; Le Galliard, J.-F.; Chabbi, A.; Greiveldinger, L.; Caquet, T.; Loreau, M.; Mougin, C.; Pichot, C.; Roy, J.; et al. How to Integrate Experimental Research Approaches in Ecological and Environmental Studies: AnaEE France as an Example. *Front. Ecol. Evol.* **2018**, *6*. [CrossRef]
5. Knapp, A.K.; Avolio, M.L.; Beier, C.; Carroll, C.J.W.; Collins, S.L.; Dukes, J.S.; Fraser, L.H.; Griffin-Nolan, R.J.; Hoover, D.L.; Jentsch, A.; et al. Pushing precipitation to the extremes in distributed experiments: recommendations for simulating wet and dry years. *Glob. Change Biol.* **2017**, *23*, 1774–1782. [CrossRef] [PubMed]
6. Ónodi, G.; Botta-Dukát, Z.; Kröel-Dulay, G.; Lellei-Kovács, E.; Kertész, M. Reduction in primary production followed by rapid recovery of plant biomass in response to repeated mid-season droughts in a semiarid shrubland. *Plant Ecol.* **2018**, *219*, 517–526. [CrossRef]
7. Peterson, D.L.; Waring, R.H. Overview of the Oregon Transect Ecosystem Research Project. *Ecol. Appl.* **1994**, *4*, 211–225. [CrossRef]

8. Li, X.; Liu, S.; Xiao, Q.; Ma, M.; Jin, R.; Che, T.; Wang, W.; Hu, X.; Xu, Z.; Wen, J.; et al. A multiscale dataset for understanding complex eco-hydrological processes in a heterogeneous oasis system. *Sci. Data* **2017**, *4*, 170083. [CrossRef] [PubMed]

9. Gardner, R.H.; Kemp, W.M.; Kennedy, V.S.; Petersen, J.E. *Scaling Relations in Experimental Ecology*; Columbia University Press: New York, NY, USA, 2012; ISBN 978-0-231-52904-4.

10. Díaz-Delgado, R.; Hurford, C.; Lucas, R. Introducing the Book "The Roles of Remote Sensing in Nature Conservation.". In *The Roles of Remote Sensing in Nature Conservation*; Springer: Cham, Switzerland, 2017; pp. 3–10, ISBN 978-3-319-64330-4.

11. Hufkens, K.; Bogaert, J.; Dong, Q.H.; Lu, L.; Huang, C.L.; Ma, M.G.; Che, T.; Li, X.; Veroustraete, F.; Ceulemans, R. Impacts and uncertainties of upscaling of remote-sensing data validation for a semi-arid woodland. *J. Arid Environ.* **2008**, *72*, 1490–1505. [CrossRef]

12. Porcar-Castell, A.; Mac Arthur, A.; Rossini, M.; Eklundh, L.; Pacheco-Labrador, J.; Anderson, K.; Balzarolo, M.; Martín, M.P.; Jin, H.; Tomelleri, E.; et al. EUROSPEC: at the interface between remote-sensing and ecosystem CO_2 flux measurements in Europe. *Biogeosciences* **2015**, *12*, 6103–6124. [CrossRef]

13. Zarco-Tejada, P.J.; Miller, J.R.; Noland, T.L.; Mohammed, G.H.; Sampson, P.H. Scaling-up and model inversion methods with narrowband optical indices for chlorophyll content estimation in closed forest canopies with hyperspectral data. *IEEE Trans. Geosci. Remote Sens.* **2001**, *39*, 1491–1507. [CrossRef]

14. Lucas, R.; Díaz-Delgado, R.; Hurford, C. Expected Advances in a Rapidly Developing Work Area. In *The Roles of Remote Sensing in Nature Conservation*; Springer: Cham, Switzerland, 2017; pp. 309–318, ISBN 978-3-319-64330-4.

15. Duan, T.; Chapman, S.C.; Guo, Y.; Zheng, B. Dynamic monitoring of NDVI in wheat agronomy and breeding trials using an unmanned aerial vehicle. *Field Crops Res.* **2017**, *210*, 71–80. [CrossRef]

16. Su, J.; Liu, C.; Coombes, M.; Hu, X.; Wang, C.; Xu, X.; Li, Q.; Guo, L.; Chen, W.-H. Wheat yellow rust monitoring by learning from multispectral UAV aerial imagery. *Comput. Electron. Agric.* **2018**, *155*, 157–166. [CrossRef]

17. Zhang, C.; Kovacs, J.M. The application of small unmanned aerial systems for precision agriculture: a review. *Precis. Agric.* **2012**, *13*, 693–712. [CrossRef]

18. Manfreda, S.; McCabe, M.; Miller, P.; Lucas, R.; Pajuelo Madrigal, V.; Mallinis, G.; Ben Dor, E.; Helman, D.; Estes, L.; Ciraolo, G.; et al. On the Use of Unmanned Aerial Systems for Environmental Monitoring. *Remote Sens.* **2018**, *10*, 641. [CrossRef]

19. Sha, Z.; Wang, Y.; Bai, Y.; Zhao, Y.; Jin, H.; Na, Y.; Meng, X. Comparison of leaf area index inversion for grassland vegetation through remotely sensed spectra by unmanned aerial vehicle and field-based spectroradiometer. *J. Plant Ecol.* **2018**. [CrossRef]

20. Vanden Borre, J.; Paelinckx, D.; Mücher, C.A.; Kooistra, L.; Haest, B.; De Blust, G.; Schmidt, A.M. Integrating remote sensing in Natura 2000 habitat monitoring: Prospects on the way forward. *J. Nat. Conserv.* **2011**, *19*, 116–125. [CrossRef]

21. Lloret, F.; de la Riva, E.G.; Pérez-Ramos, I.M.; Marañón, T.; Saura-Mas, S.; Díaz-Delgado, R.; Villar, R. Climatic events inducing die-off in Mediterranean shrublands: are species' responses related to their functional traits? *Oecologia* **2016**, *180*, 1–13. [CrossRef] [PubMed]

22. Ciais, P.; Reichstein, M.; Viovy, N.; Granier, A.; Ogée, J.; Allard, V.; Aubinet, M.; Buchmann, N.; Bernhofer, C.; Carrara, A.; et al. Europe-wide reduction in primary productivity caused by the heat and drought in 2003. *Nature* **2005**, *437*, 529–533. [CrossRef]

23. de la Riva, E.G.; Lloret, F.; Pérez-Ramos, I.M.; Marañón, T.; Saura-Mas, S.; Díaz-Delgado, R.; Villar, R. The importance of functional diversity on the stability of Mediterranean shrubland communities after the impact of extreme climatic events. *J. Plant Ecol.* **2017**, *10*, 281–293. [CrossRef]

24. Kröel-Dulay, G.; Ransijn, J.; Schmidt, I.K.; Beier, C.; De Angelis, P.; de Dato, G.; Dukes, J.S.; Emmett, B.; Estiarte, M.; Garadnai, J.; et al. Increased sensitivity to climate change in disturbed ecosystems. *Nat. Commun.* **2015**, *6*, 6682. [CrossRef] [PubMed]

25. Lellei-Kovács, E.; Kovács-Láng, E.; Kalapos, T.; Botta-Dukát, Z.; Barabás, S.; Beier, C. Experimental warming does not enhance soil respiration in a semiarid temperate forest-steppe ecosystem. *Community Ecol.* **2008**, *9*, 29–37. [CrossRef]

26. Rouse, J.W. *Monitoring the Vernal Advancement and Retrogradation (Green Wave Effect) of Natural Vegetation*; Texas A&M Univ.; Remote Sensing Center: College Station, TX, USA, 1974.

27. Ónodi, G.; Kröel-Dulay, G.; Kovács-Láng, E.; Ódor, P.; Botta-Dukat, Z.; Lhotsky, B.; Barabás, S.; Garadnai, J.; Kertész, M. Comparing the accuracy of three non-destructive methods in estimating aboveground plant biomass. *Community Ecol.* **2017**, *18*, 56–62. [CrossRef]

28. Molnár, Z.; Biró, M.; Bartha, S.; Fekete, G. Past Trends, Present State and Future Prospects of Hungarian Forest-Steppes. In *Eurasian Steppes. Ecological Problems and Livelihoods in a Changing World*; Werger, M.J.A., van Staalduinen, M.A., Eds.; Plant and Vegetation; Springer Netherlands: Dordrecht, The Netherlands, 2012; pp. 209–252, ISBN 978-94-007-3886-7.

29. Kovács-Láng, E.; Kröel-Dulay, G.; Kertész, M.; Fekete, G.; Bartha, S.; Mika, J.; Dobi-Wantuch, I.; Rédei, T.; Rajkai, K.; Hahn, I. Changes in the composition of sand grasslands along a climatic gradient in Hungary and implications for climate change. *Phytocoenologia* **2000**, *30*, 385–407. [CrossRef]

30. Bartholy, J.; Pongrácz, R. Regional analysis of extreme temperature and precipitation indices for the Carpathian Basin from 1946 to 2001. *Glob. Planet. Change* **2007**, *57*, 83–95. [CrossRef]

31. Beier, C.; Emmett, B.; Gundersen, P.; Tietema, A.; Peñuelas, J.; Estiarte, M.; Gordon, C.; Gorissen, A.; Llorens, L.; Roda, F.; et al. Novel Approaches to Study Climate Change Effects on Terrestrial Ecosystems in the Field: Drought and Passive Nighttime Warming. *Ecosystems* **2004**, *7*, 583–597. [CrossRef]

32. Peñuelas, J.; Prieto, P.; Beier, C.; Cesaraccio, C.; Angelis, P.D.; Dato, G.D.; Emmett, B.A.; Estiarte, M.; Garadnai, J.; Gorissen, A.; et al. Response of plant species richness and primary productivity in shrublands along a north–south gradient in Europe to seven years of experimental warming and drought: reductions in primary productivity in the heat and drought year of 2003. *Glob. Change Biol.* **2007**, *13*, 2563–2581. [CrossRef]

33. Díaz-Delgado, R.; Lloret, F.; Pons, X.; Terradas, J. Satellite Evidence of Decreasing Resilience in Mediterranean Plant Communities After Recurrent Wildfires. *Ecology* **2002**, *83*, 2293–2303. [CrossRef]

34. Haberl, H.; Winiwarter, V.; Andersson, K.; Ayres, R.U.; Boone, C.; Castillo, A.; Cunfer, G.; Fischer-Kowalski, M.; Freudenburg, W.R.; Furman, E.; et al. From LTER to LTSER: Conceptualizing the socioeconomic dimension of long-term socioecological research. *Ecol. Soc.* **2006**, *11*, 13. [CrossRef]

35. Mirtl, M.; Orenstein, D.E.; Wildenberg, M.; Peterseil, J.; Frenzel, M. *Development of LTSER Platforms in LTER-Europe: Challenges and Experiences in Implementing Place-Based Long-Term Socio-ecological Research in Selected Regions*; Springer Netherlands: Dordrecht, The Netherlands, 2013; ISBN 978-94-007-1176-1.

36. Dick, J.; Orenstein, D.E.; Holzer, J.; Wohner, C.; Achard, A.-L.; Andrews, C.; Avriel-Avni, N.; Beja, P.; Blond, N.; Cabello, J.; et al. What is socio-ecological research delivering? A literature survey across 25 international LTSER platforms. *Sci. Total Environ.* **2018**, *622–623*, 1225–1240. [CrossRef]

37. Franklin, S.E.; Ahmed, O.S.; Williams, G. Northern Conifer Forest Species Classification Using Multispectral Data Acquired from an Unmanned Aerial Vehicle. *Photogramm. Eng. Remote Sens.* **2017**, *83*, 501–507. [CrossRef]

38. Shen, Y.-Y.; Cattau, M.; Borenstein, S.; Weibel, D.; Frew, E.W. Toward an Architecture for Subalpine Forest Health Monitoring Using Commercial Off-the-Shelf Unmanned Aircraft Systems and Sensors. Proceedings of 17th AIAA Aviation Technology, Integration, and Operations Conference, Denver, CO, USA, 5–9 June 2017.

39. Padró, J.-C.; Carabassa, V.; Balagué, J.; Brotons, L.; Alcañiz, J.M.; Pons, X. Monitoring opencast mine restorations using Unmanned Aerial System (UAS) imagery. *Sci. Total Environ.* **2019**, *657*, 1602–1614. [CrossRef]

40. Hakala, T.; Markelin, L.; Honkavaara, E.; Scott, B.; Theocharous, T.; Nevalainen, O.; Näsi, R.; Suomalainen, J.; Viljanen, N.; Greenwell, C.; et al. Direct Reflectance Measurements from Drones: Sensor Absolute Radiometric Calibration and System Tests for Forest Reflectance Characterization. *Sensors* **2018**, *18*, 1417. [CrossRef] [PubMed]

41. Ahmed, O.S.; Shemrock, A.; Chabot, D.; Dillon, C.; Williams, G.; Wasson, R.; Franklin, S.E. Hierarchical land cover and vegetation classification using multispectral data acquired from an unmanned aerial vehicle. *Int. J. Remote Sens.* **2017**, *38*, 2037–2052. [CrossRef]

42. Unger, J.; Reich, M.; Heipke, C. UAV-based photogrammetry: monitoring of a building zone. *Int. Arch. Photogramm. Remote Sens. Spat. Inf. Sci.* **2014**, *XL*, 601–606. [CrossRef]

43. Rasmussen, J.; Ntakos, G.; Nielsen, J.; Svensgaard, J.; Poulsen, R.N.; Christensen, S. Are vegetation indices derived from consumer-grade cameras mounted on UAVs sufficiently reliable for assessing experimental plots? *Eur. J. Agron.* **2016**, *74*, 75–92. [CrossRef]

44. Lu, B.; He, Y.; Liu, H.H.T. Mapping vegetation biophysical and biochemical properties using unmanned aerial vehicles-acquired imagery. *Int. J. Remote Sens.* **2017**, *39*, 1–23. [CrossRef]

45. Levin, N.; Lugassi, R.; Ramon, U.; Braun, O.; Ben-Dor, E. Remote sensing as a tool for monitoring plasticulture in agricultural landscapes. *Int. J. Remote Sens.* **2007**, *28*, 183–202. [CrossRef]
46. Jiménez, M.; Díaz-Delgado, R. Sub-pixel Mapping of Doñana Shrubland Species. In *The Roles of Remote Sensing in Nature Conservation*; Springer: Cham, Switzerland, 2017; pp. 141–163, ISBN 978-3-319-64330-4.

drones

MDPI

Article

Greenness Indices from a Low-Cost UAV Imagery as Tools for Monitoring Post-Fire Forest Recovery

Asier R. Larrinaga [1,2,3,*] and Lluis Brotons [1,4,5]

1   InForest Joint Research Unit, (CTFC-CREAF), 25280 Solsona, Spain; lluis.brotons@ctfc.cat
2   eNeBaDa, Santiago de Compostela, 15892 A Coruña, Spain
3   Forest Genetics and Ecology Group, Biologic Mission of Galicia (CSIC), 36413 Pontevedra, Spain
4   CREAF, 08193 Cerdanyola del Vallès, Spain
5   CSIC, 08193 Cerdanyola del Vallès, Spain
*   Correspondence: asier@enebada.eu or arodriguez@mbg.csic.es; Tel.: +34-659-087-889

Received: 1 November 2018; Accepted: 1 January 2019; Published: 6 January 2019

Abstract: During recent years unmanned aerial vehicles (UAVs) have been increasingly used for research and application in both agriculture and forestry. Nevertheless, most of this work has been devoted to improving accuracy and explanatory power, often at the cost of usability and affordability. We tested a low-cost UAV and a simple workflow to apply four different greenness indices to the monitoring of pine (*Pinus sylvestris* and *P. nigra*) post-fire regeneration in a Mediterranean forest. We selected two sites and measured all pines within a pre-selected plot. Winter flights were carried out at each of the sites, at two flight heights (50 and 120 m). Automatically normalized images entered an structure from motion (SfM) based photogrammetric software for restitution, and the obtained point cloud and orthomosaic processed to get a canopy height model and four different greenness indices. The sum of pine diameter at breast height (DBH) was regressed on summary statistics of greenness indices and the canopy height model. Excess green index (ExGI) and green chromatic coordinate (GCC) index outperformed the visible atmospherically resistant index (VARI) and green red vegetation index (GRVI) in estimating pine DBH, while canopy height slightly improved the models. Flight height did not severely affect model performance. Our results show that low cost UAVs may improve forest monitoring after disturbance, even in those habitats and situations where resource limitation is an issue.

Keywords: low-cost UAV; greenness index; *Pinus nigra*; *Pinus sylvestris*; forest regeneration; flight altitude; small UAV

1. Introduction

During recent years, UAVs (unmanned aerial vehicles) have grown increasingly popular for the study of land and its cover [1,2]. This trend is the consequence of a recent exponential development of both the UAV industry and the do-it-yourself community, fostered by the technological advances in robotics and the miniaturization of electronics.

Forest research is one of the fields where the use of UAV has promised immediate benefits [1]. UAVs allow the reduction of costs of airborne photography and LIDAR, and approach technology to its final user [2,3]. By doing so, it offers great flexibility. UAV deployment is fast and cheap, ensuring rapid responses to the needs of both academia and industry while allowing for repeated sampling with no limits on deployment periodicity. A new, tailor-cut telemetry sampling strategy is now possible, designed to fit the specific needs of each case study [3].

UAVs have been used to discriminate among species and provide estimates of tree and stand size, tree cover, canopy height, gap abundance or even productivity, alone or in combination with LIDAR data [4–12]. Tree health and pathogen or parasite attack have also been evaluated by means of UAV telemetry [13].

Therefore, research on different methodologies and, more specifically, on the accuracy of those data estimated by means of UAV imagery has bloomed during the last five years [2]. The accuracy level of these kind of works is continuously increasing and it is a major focus of an important body of research on UAV use in forestry [2,4,14–16]. Spatial accuracy and automatic tree detection in particular are rapidly improving [9,12,16–19].

Nevertheless, accuracy is not always the main constraint to UAV use in forestry science. In fact, pursuing high standards of spatial and analytical accuracy is a time consuming goal that often requires major investment [20]. While this might be a sensible approach when working with economically exploited forests and woodlands, it might hinder the development of forestry research in other areas or research fields, such as disturbance response, where immediate revenues cannot be envisaged. In these situations, the absence of any kind of data is common while economic resources and work forces are scarce. A reduction of both accuracy and the cost of deployment could hence help to get general data on the ecology of the forest that would greatly improve our knowledge.

In this work we explore the use of a low-cost UAV platform as a tool for monitoring recovery of a Mediterranean forest after a strong disturbance. Our objective is to assess post-fire pine regeneration (Scots pine, *Pinus sylvestris*, and black pine, *P. nigra*) in an area affected by a wildfire where oak has become dominant. In order to identify pine cover in our area, we compare the use of four different greenness indices at two different flight heights, recorded areas and hence costs. Low-cost UAV platforms offer several advantages for this task: (1) they are affordable and easy to use for stakeholders and practitioners, (2) they can be controlled and analyzed with free photogrammetric software and 3D reconstruction web services, (3) due to their small size they can be flown even in remote areas, where access by vehicle is difficult, (4) they can be flown on demand at no cost, allowing one to choose the flying time depending on weather, plant phenology or other logistical constraints, and (5) they offer ultra-high resolution, allowing one to detect pine trees even at the early stages of regeneration. In this context, we aimed to develop a tool to monitor the emergence of pines that grow among the oaks and test the suitability of low-cost UAVs as a cost effective monitoring tool.

2. Materials and Methods

We carried out our work in the municipality of Riner, in the Lleida province (Catalonia, Spain), where an extensive wildfire burnt down around 25,000 ha of pine-dominated woodland in 1998. Most of the area have apparently recovered to a great extent in these last 20 years. A closer look, however, depicts a different picture. The effect of wildfires might have drawn the forest beyond its resilience threshold, causing a change to a new alternative equilibrium state [21]. In fact, although tree cover seems to be almost completely recovered, its species composition has changed in a radical way, as Portuguese oak (*Quercus faginea*) thrives in the burnt land where pines (*P. nigra* and *P. sylvestris*) were dominant before the wildfire.

Our two sites, La Carral and Cal Rovira-Sanca, are 3.7 km away from each other and both of them supported similar Mediterranean forests before the 1998 fire, on marl, limestone, and sandstone rocks (Figures S1 from reference [22]).

2.1. UAV Deployment and Field Sampling

We carried out two flights in each of the sites, one at a height of 50 m and a second one at a height of 120 m over terrain, by means of a DJI Phantom 2 quadcopter. We flew at two altitudes with the aim of exploring the effect of flight height on image quality and subsequent capacity to characterize pine tree recovery. The copter was equipped with a Phantom Vision FC200 camera, manufactured by DJI, which has a resolution of 14 Mpx with a sensor size of 1/2.3″ (6.17 mm * 4.55 mm), a focal length of 5 mm and an electronic rolling shutter [23]. We set the camera to shoot one picture every three (La Carral) and five (Cal Rovira-Sanca; Table 1) seconds with an automatic exposition mode setting (with ISO 100).

All four flights were carried out in March 2015 (17 years after the fire) under optimal weather conditions (sunny days with low wind intensity); in Cal Rovira-Sanca in the morning and in the afternoon in La Carral (Table 1). By flying in winter, we ensured a good spectral discrimination between pines (the only perennial tree species group in the area)—and the remaining components of the canopy, namely Portuguese oaks, as the latter still hold their dry leaves on the branches and do not shed new leaves until spring.

In each of the sites we selected a sampling area near the center of the flight zone and identified all the pines growing there (Figure 1). Each pine was geo-referenced and identified to the species level. In addition, height and diameter at breast height was measured (DBH) for each individual. DBH was measured with the aid of a measuring tape with a 1 mm resolution. Tree height was measured with the same measuring tape, except for the highest trees, where a measuring pole was used (1 cm resolution). We measured DBH at 1 m height for logistical reasons, given the abundance of low pine trees.

Table 1. Characteristics of the four flights carried out in La Carral and Cal Rovira-Sanca. Time refers to mean time of each flight. Centre of scene: geographic center of each scene in UTM, fuse 31, datum ETRS89. Flight height gives nominal values. Area: coverage area of each flight. Pixel size: size of ground pixel. Reprojection error: difference between a point in an image and its position according to the fitted 3D model. Motion blur: blur due to linear movement (rotation effects are not included).

Site	La Carral	La Carral	Cal Rovira-Sanca	Cal Rovira-Sanca
Date (DD/MM/YY)	03/07/18	03/07/18	03/03/18	03/03/18
Time (UTC)	16:59	17:26	10:13	10:25
Sun elevation angle (°)	19.08	14.49	27.29	28.96
Sun azimuth angle (°)	243.95	249.07	130.07	132.86
Centre of Scene (UTM31N-ETRS89)	(378698,4640314)	(378715,4640324)	(375511,4642336)	(375522,4642349)
# of images	160	147	90	67
Flight height (m)	50	120	50	120
Flight speed (m/s)	4	4	4	4
Area (ha)	5.82	24.6	7.54	21.3
Side overlap (%)	55	65	48	62
Forward overlap (%)	74	89	57	82
Effective overlap (# image/pixel)	3.40	7.94	2.88	4.80
Pixel size (cm)	1.46	4	1.59	3.96
Reprojection error (pixel)	8.31	7.95	1.78	1.76
Mean shutter speed (s)	1/288	1/312	1/457	1/525
Motion blur (cm - pixel)	1.39–0.95	1.28–0.32	0.88–0.55	0.76–0.16

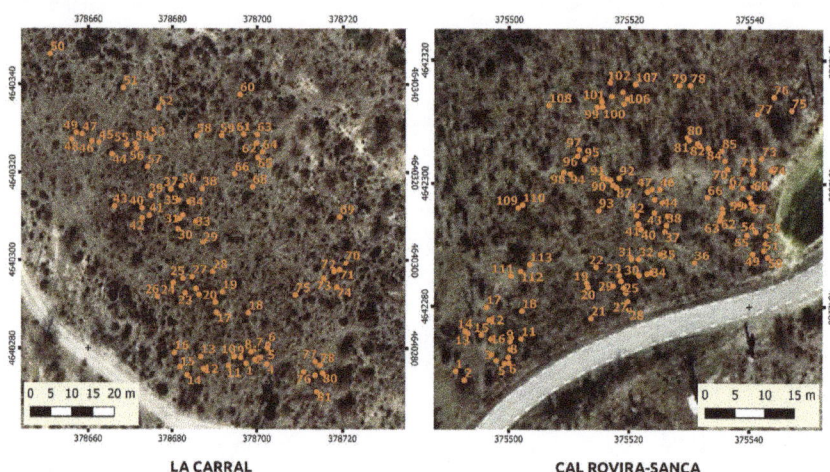

LA CARRAL **CAL ROVIRA-SANCA**

Figure 1. Aerial ortophotographs of both sites, with overlapped positions of all the pines found in the sampling areas.

2.2. Image Analysis

We aimed to get an easy workflow and to that end, we tried to minimize parameter tweaking during the whole process. Hence, most of the procedures were carried out using automatic tools and algorithms, as well as default values for the various parameters. For the rest of the section, we will be giving details on the parameters we set specifically, while omitting those parameters that were set to their default values.

First, we adjusted all the aerial images with an image managing and editing software [24], by sequentially applying the Automatic Levels and Automatic Contrast tools. By doing so we corrected some inequalities in color balance produced by the FC200 camera and improved their contrast for further analysis.

We carried out 3D reconstruction within an advanced terrain-oriented software that creates 3D models, point clouds and orthomosaics from a set of images on the same subject, by means of SfM (structure from motion) techniques [16,25,26]. We followed the general workflow suggested in its user manual [27]. Photo aligning and estimation of camera locations and calibration parameters were carried out as a first step, together with the building of a sparse point cloud (Figure 2; Figure S2). This sparse point cloud was then used to create a dense point cloud with mean point density of 407 points/m for the 50 m high flights and 69 points/m for the 120 m high flights. Finally, a digital elevation model and an orthomosaic were built from this point cloud, using the corresponding default parameters in our 3D reconstruction software [25].

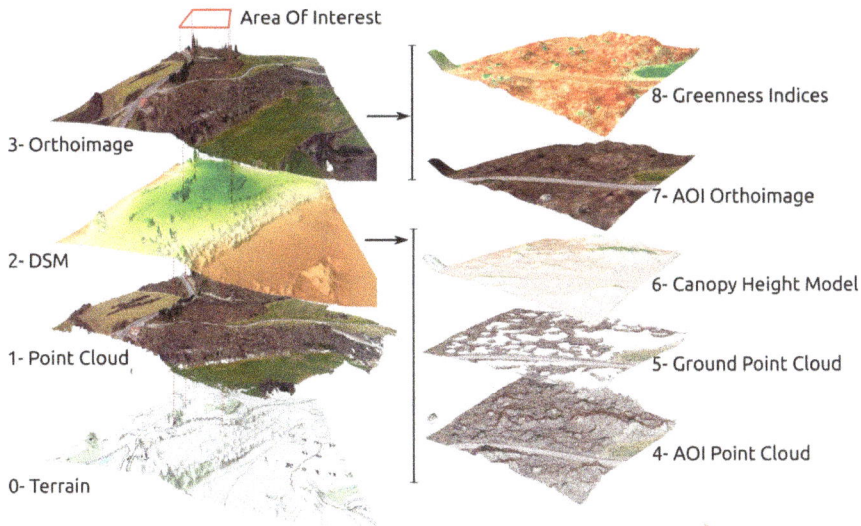

Figure 2. General workflow for the analysis of UAV imagery, as shown by its intermediate output images, point clouds and 3D models. A detailed workflow is provided as Figure S2-1. Orthoimages are shown draped on the obtained DSM, in order to improve figure display.

As a reference image to set control points, we used the most recent orthophoto of the area from the National Plan of Aerial Orthophotography (Plan Nacional de Ortofotografía Aerea, PNOA), which has a pixel resolution of 25 cm. Altitude of control points was manually extracted from the official digital terrain model (DTM05, 5 m spatial resolution) released by PNOA. We set between 7 and 9 control points per flight by selecting particular features that could be identified both in the orthomosaic obtained from the 3D reconstruction software [25] and in our reference aerial photograph, such as stones, rocks, fallen trunks and artificial features (corners, road paint, etc.). We then used those control points to reconfigure the cameras and rebuild the dense point cloud, the DSM (digital surface model) and the orthomosaic.

The entire process within the 3D reconstruction software was carried out with by-default parameters and pixel size was only set during the export step. We exported 50 m high flight orthomosaics and DSM with a pixel size of 2 and 8 cm, respectively and those from 120 m high flights with pixels of 4 and 16 cm.

We clipped the obtained point cloud by means of a specific free tool for forest LIDAR data analysis [28]). Misplaced points were detected in those clipped point clouds obtained from 50 m high flights and hence, they were cleaned up with a point cloud editing and registering open software [29]. Those from both 120 m high flights did not need such a cleaning process. We extracted those points corresponding to the ground level, by means of a filtering algorithm adapted from Kraus and Pfeifer [30] and then computed a digital terrain model (DTM) by using the mean elevation of all ground points within a cell [28]. From the highest elevation point in each cell and the created DSM, we built a canopy height model with a cell size of 10*10 cm^2 [28].

We calculated four different greenness indices from the obtained orthomosaic: excess green index (ExGI), green chromatic coordinate (GCC), green-red vegetation index (GRVI) and visible atmospherically resistant index (VARI). We hereby provide the rationale for their selection and their definition:

- The excess green index (ExGI) is a contrast index that has been shown to outperform other indices in discriminating vegetation [31,32] and is one of the most widely used indices in the visual spectrum. It is defined as:

$$ExGI = 2 * G - (R + B) \tag{1}$$

- The green chromatic coordinate (GCC) has also been used to detect vegetation and analyze plant phenology and dynamics [31,32]. Both ExGI and GRVI correlate with measurements made with a SpectroSense narrow spectrometer [33], but GRVI is far less sensitive to changes in scene illumination [32]. It is simply the chromatic coordinate of the green channel expressed as a proportion of the sum of coordinates:

$$GCC = \frac{G}{R + G + B} \tag{2}$$

- The green red vegetation index (GRVI) was first used by Rouse et al. [34] who concluded it could be used for several measures of crops and rangelands. Their conclusions have been later confirmed in several occasions [35–38]. It responds to leave senescence of deciduous forests in a parallel way to that of NDVI [37] and hence could be useful for discriminating senescent leaves from green needles. This index is given by:

$$GRVI = \frac{G - R}{G + R} \tag{3}$$

- Lastly, the visible atmospherically resistant index (VARI) was proposed by Gitelson et al. [39]. It is an improvement of GRVI that reduces atmospheric effects. Although this is not an expected severe effect in low flying UAV platforms, it might locally be so, at Mediterranean sites with large amounts of bare soil. In addition, it has been reported to correlate better than GRVI with vegetation fraction [39]. It is defined as:

$$VARI = \frac{G - R}{G + R - B} \tag{4}$$

We calculated all four greenness indices directly from their digital numbers (DN) as provided by the JPEG format provided by the camera, instead of calculating reflectance values. JPEG compression is a "lossy" compression method that prioritizes brightness over color, resulting in an important reduction of dynamic range of the picture and certain degree of image posterization (hard color changes and lower number of colors in the image). As a result, radiometric resolution decreases, which should reduce the ability to discriminate among different terrain or vegetation categories based on their visual spectrum. However, JPEG images still can successfully discriminate among different phenological stages of the vegetation, except at the most extreme compression ratios (97%) [32].

As we were not using reflectance to calculate these indices, their properties might not be the same as those described by other authors [35,37,39]. Particularly, our calculated indices cannot be directly compared with indices from different studies or even different flights, as they are sensitive to sensor characteristics and scene illumination. Still, we expected them to be useful in our context and decided to use this simpler approach, as our aim was to get as simple a workflow as possible, in order to allow for an easy, handy use of UAV imagery by non-expert users. Similar approaches have been successful in the past, even when analyzing repeated images over time [32,33,40,41]. Despite important effects of camera model and scene illumination in absolute greenness indices, the changes in plant phenology (changes from green to senescent leaves) were correctly detected by using uncorrected DNs [41].

Index calculations and their posterior analysis were carried out in an open SIG software [42], by combining the use of raster calculator with specific tools of zone statistics and spatial joining. The whole process for each index, after its calculation, involved the following steps:

- Applying the greenness threshold to the indices layers, in order to erase all non-green pixels, which were set to 0, For ExGI, GRVI and VARI, we defined green pixels as those with values greater than 0 and reclassified values less than 0 as 0. For GCC, the applied threshold was 1/3, and hence all values equal to or lesser than 1/3 were set to 0.
- Calculating the zone statistics of the greenness index and the filtered canopy height model for each 5*5 m^2 cell within the study area. Zone statistics produces six different measures of the index value per each cell: count, mean, standard deviation, median, maximum and minimum.
- Calculating the pooled DBH of all measured pines within the cells of this same grid.

2.3. Statistical Analysis

We aimed to assess the recovery of pines in the areas burnt in 1998. Hence, we selected the sum of the diameter at the breast height (DBH) of all pines within a cell as our response variable (Figure 3). DBH was highly correlated to pine height (see Supplementary Material S1) and its measure in pines is easier and less prone to measurement errors. DBH has also been related to many other morphological and functional traits [43–49] and hence it is open to a more insightful analysis. The sum of DBH values from all the pines in a grid cell combines the effect of density (number of pine trees per cell) with that of tree size (DBH). Given the relationship of DBH with crown size and foliage area [50], the sum of DBH is expected to correlate also with canopy cover [51]. In fact, basal area and tree density can successfully be used to predict canopy cover in pines [52–54].

First, we carried out simple linear regression analysis of the sum of DBH on the four greenness indices and on the canopy height model. For each greenness index and the canopy height model, we fitted six models, where the derived explanatory variables corresponded to six different summary statistics per 5 m grid-cell: the count of points with non-zero values for the corresponding index or canopy height model and the sum, mean, standard deviation, maximum and median of the values of all points within the cell. Our aim was to explore which statistic could be better suited for estimating sum of DBH from greenness index and canopy height maps. Then we tried to improve model fit by combining greenness indices with pine canopy height, by means of multiple linear regression analysis.

R^2 (for simple linear models) and adjusted R^2 (for multiple linear models) and scatterplots were used throughout the process to assess model fit and comparing models.

All statistics were carried out in R [55], by means of the R-Studio integrated development environment [56].

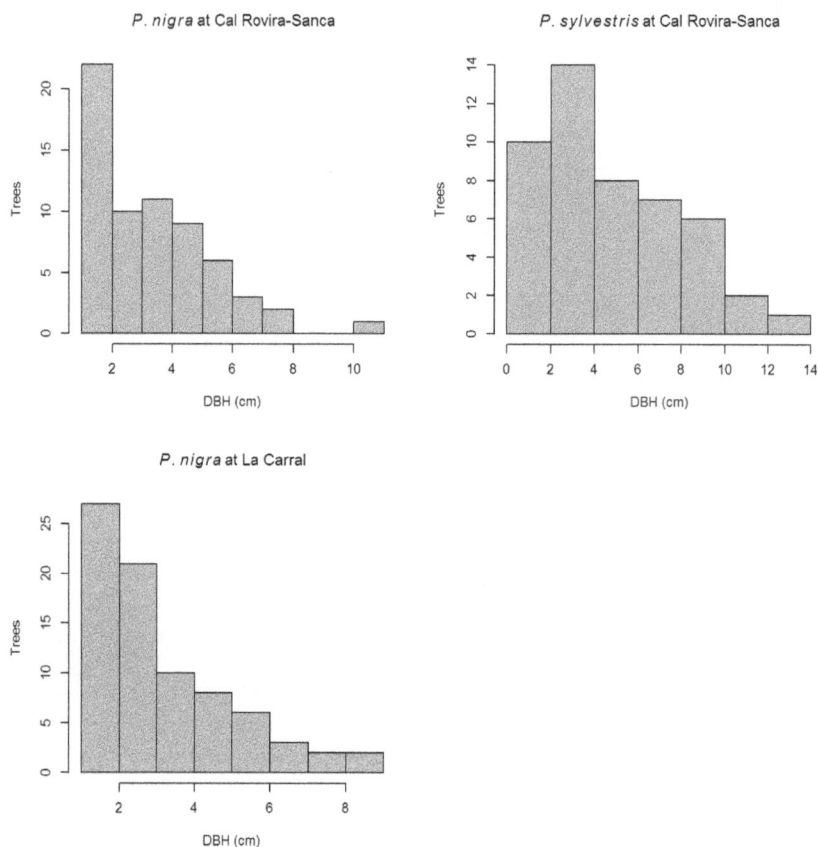

Figure 3. Distribution of DBH for all individuals of *Pinus nigra* and *Pinus sylvestris* found within the study area, both at Cal Rovira-Sanca and La Carral.

3. Results

Two pine species were found in both study sites, black pine (*Pinus nigra*) and Scots pine (*Pinus sylvestris*), although only one individual of the latter was recorded in La Carral. Mean tree height was 124.94 cm (s.d. = 88.78) for black pine and 188.15 (s.d. = 117.87) for Scots pine in Cal Rovira-Sanca and 109.82 (s.d. = 51.38) for black pine in La Carral, while the only Scots pine in this site was 230 cm high. DBH values were lower for black pine than for Scots pine, around 3.25 cm (mean = 3.36 and s.d. = 2.1 for Cal Rovira-Sanca; mean = 3.2 and s.d. = 1.87 for La Carral) versus 5 (mean = 4.91 and s.d. = 3.12 in Cal Rovira-Sanca; 6.6 cm for the only tree in La Carral).

The proportion of small trees was high for both species, but higher for black pine: median height of 117.5 and 105 and median DBH of 3.1 and 2.4 for Cal Rovira-Sanca and La Carral, respectively, as opposed to a median height of 192.5 and median DBH of 4.1 in Cal Rovira-Sanca for the Scots pine (see Figure 3 and Figure S3).

Due to flying times, both at La Carral and Cal Rovira-Sanca, light availability strongly limited the automatic selection of shutter speed, which resulted in a high motion blur (Table 1). Still, it was much higher in La Carral than in Cal Rovira-Sanca, which translated into additional distortion and reprojection errors (Table 1, Figure S4). Reprojection error was more than four times larger in La Carral than in Cal Rovira-Sanca and radial distortion much more exaggerated.

Greenness indices were revealed to be much better proxies for the sum of pine DBH than canopy height in both La Carral and Cal Rovira-Sanca. In fact, mean R^2 for the regressions of sum of pine DBH on the different statistics of greenness indices ranged from 0.005 to 0.466, while the regressions on canopy statistics ranged from 0.008 to 0.028 (Table 1; Figure 4). However, large differences were found among greenness indices (Figure S5-1). Even if GRVI and VARI still resulted in higher R^2 than any pine canopy height model, they were clearly outperformed by ExGI and GCC, which reached a maximum R^2 of more than 45% (Table 2).

Different statistics for the four greenness indices showed markedly differing fits, as shown by the high coefficients of variation of R^2 (Table 1; Figure 4). There were no clear-cut patterns of precision among the tested statistics that could be considered general for all four indices. Nevertheless, the count of non-zero values appears as the most consistently unreliable measure, with CVs higher than 100% for the all four indices. The best fitting model for one flight was not the same as for the others (Tables S6), which precludes considering any of them as a general best fitting model (Figure S7).

Table 2. Summary statistics of the determination coefficient (R^2) for the simple regressions of the sum of pine DBH on the four greenness indices and the canopy height model. Figures show the average R^2 across flights (two sites at each of two flight heights) and its coefficient of variation, expressed as percentage (within brackets). CHM stands for canopy height model. See "Imagery analysis" for a definition of the four indices.

	ExGI	GRVI	GCC	VARI	All Indices	CHM
Count of non-zero index	0.154 (124.0)	0.114 (118.4)	0.147 (127.9)	0.121 (119.8)	0.134 (14.5)	0.028 (50.0)
Max index value	0.389 (27.2)	0.164 (98.2)	0.297 (37.7)	0.051 (119.6)	0.225 (65.9)	0.014 (121.4)
Mean index value	0.401 (12.7)	0.195 (61.0)	0.370 (50.5)	0.039 (184.6)	0.251 (66.9)	0.015 (113.3)
Median index value	0.227 (80.6)	0.162 (50.6)	0.182 (103.3)	0.076 (118.4)	0.162 (39.1)	0.008 (125.0)
Std index values	0.440 (37.3)	0.152 (57.9)	0.466 (32.2)	0.005 (40.0)	0.266 (84.5)	0.018 (100.0)
Sum of index values	0.256 (40.2)	0.088 (78.4)	0.155 (122.6)	0.027 (122.2)	0.132 (74.6)	0.014 (114.3)
All measures	0.311 (36.8)	0.146 (26.4)	0.270 (48.4)	0.053 (76.8)		0.016 (41.1)

Canopy height alone shows a very poor explanatory power of the sum of DBH of pines, with a pooled mean of R^2 of 0.016, well below the values obtained from the four greenness indices. Accordingly, jointly considering greenness indices and canopy height does not increase adjusted R^2 more than 10% (Figure 4). For both sites and both flight heights, however, the best fitting model included always a canopy height statistics (Figure 5; Tables S6).

Different statistics of the canopy height model resulted in markedly different models (Table 1), with adjusted R^2s varying between 0.8 and 28%. The canopy statistics that provided the highest R^2 was the count of cells with non-zero values, although it was not the one producing the best models when combined with greenness indices (Table S1).

Flight height did not severely affect the capacity to estimate the sum of pine DBH per grid-cell (Figures 4 and 5). However, the best fits were always achieved for 120 m flights (Figure 4), although with a larger difference in Cal Rovira-Sanca.

Overall, the model with a best fit for the four flights included the standard deviation of the GCC index and the median of the canopy height model. It generally achieves fits close to those of the best fitting model for each of the flights (Figure 4 and Figure S5-1, Tables S3), although the difference was higher for the 50 m flight in Cal Rovira-Sanca.

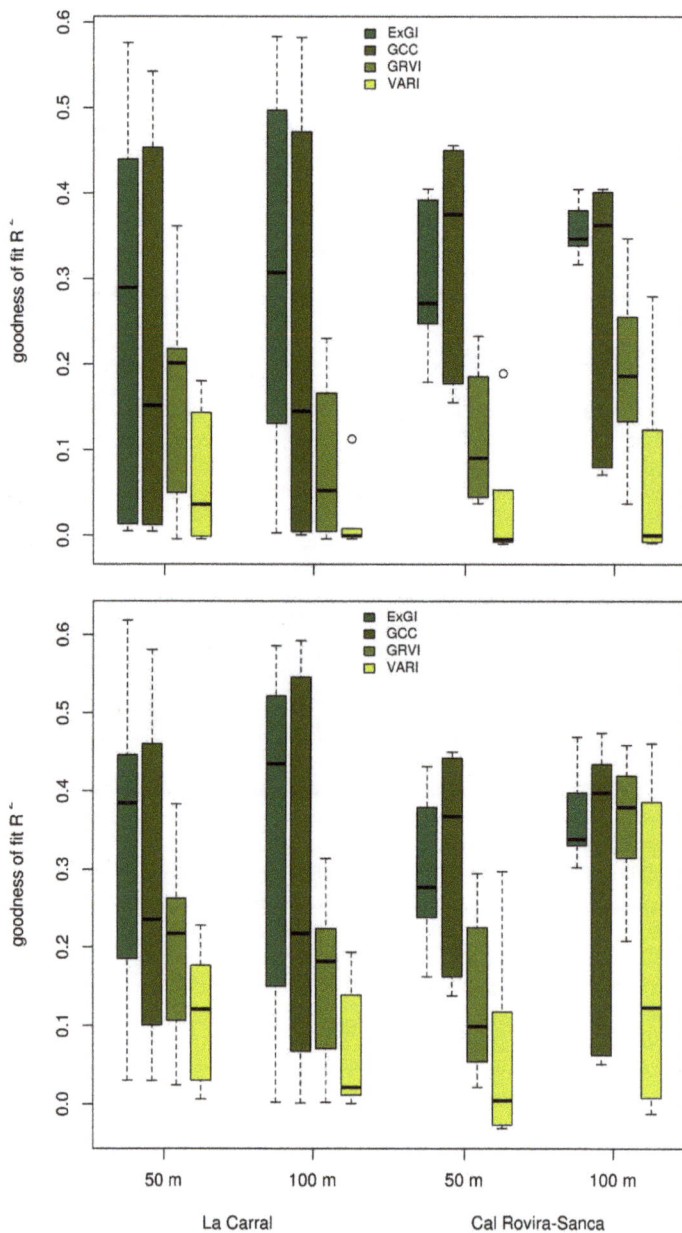

Figure 4. Values of the determination coefficient from the OLS regressions of the four greenness indices on the sum of DBH. Upper panel: simple regressions. Lower panel: multiple regressions where canopy height statistics were included as additional explanatory variables.

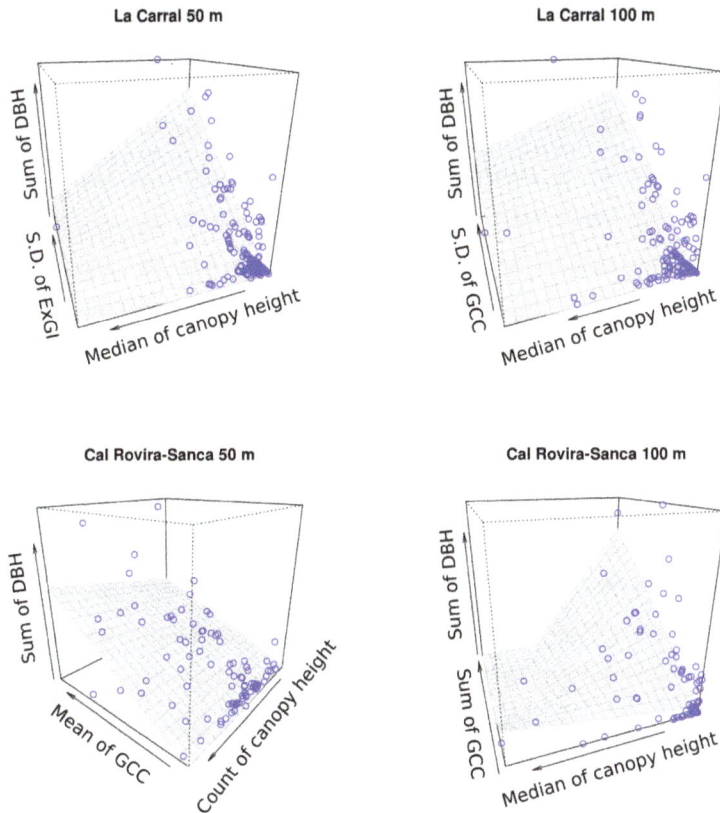

Figure 5. 3D scatterplots showing the best models. Each panel shows the bivariate regression that obtained the best fit (highest R^2) at each of the four flights. Blue dots indicate observed values, while the grey cover shows the fitted bivariate regression surface. Sum of DBH refers to the sum of the DBH of all the trees in each of the 5 m cells in which the study area was divided. Mean, s.d. and sum of GCC refers to the mean, standard deviation and sum of the Green Chromatic Coordinate values for all points of the point cloud within those same cells. S.d. of ExGi refers to the standard deviation of the Excess Green Index values for these same cells. Finally, median of canopy height model refers to the median value of the canopy height model (digital surface model) in each of the 5 m cells.

4. Discussion

UAVs are being increasingly used in ecological and conservation studies and monitoring, due to their multiple advantages, such as an increased spatial resolution, lower cost, higher acquisition flexibility and higher temporal resolution [1,2]. Low cost UAVs increase flexibility of use and reduce general costs as compared to professional platforms, and at the same time considerably reduce their learning slope, which has been deemed too steep. Yet, despite their low cost and ease of use, they accomplish spatial resolution way higher than those of aerial orthoimagery and provide useful results with a limited effort in image post-processing and analysis. This cost reduction is especially important for ecological studies and monitoring in non-profit forests and other habitats. In fact, nowadays, close forests make up only a small fraction of the experiments with UAVs in the field, while high value crops and forests are the main focus of most of them [6].

The flexibility and low cost of UAV deployment have been key factors determining the results of our pine estimation, as they allowed flying during the end of winter, before Portuguese oaks shed their leaves and the grass begins growing after the winter. Seeking the best moment to maximize spectral contrast among habitat components is paramount when it comes to discriminating different plant species or vegetation types with visual spectrum cameras [40].

Since the proposal of the first two vegetation indices by Pearson and Miller [57] and that of the NDVI shortly after by Rouse et al. [34], a high number of vegetation indices have been proposed and used for the study of vegetation from satellite, airplane and UAV sensor imagery (see for example references [32,38,39,41,58,59]). Most of them rely on the use of near infrared region (NIR), which profits from the marked difference in the absorbance spectrum of chlorophyll between red and near-infrared regions. The use and development of those indices that rely exclusively on the visual spectrum has sharply increased during last years, though, as a way to overcome the limited choice of spectrum bands in most UAV carried sensors. Assessment of their differential performance has yielded inconclusive results [31,32,35–37,39,41], but may even be better than those that include a NIR band [33,36]. Our results match those of Woebbecke et al. [31], who reported ExGI and GCC to perform better than other four indices when discriminating vegetation from non-plant background in different light conditions. Sonnentag et al. [32] recommended using GCC over ExGI to examine vegetation phenology in repeat photography studies, due to its lower sensitivity to changes in scene illumination. In our two sites, GCC also achieved higher accuracies when estimating pine cover in regenerating forests than ExGI.

By combining these indices with information derived from a canopy height model, we obtained determination coefficients for the pooled DBH of pines of up to 60%. Similar studies in agricultural plots, relating vegetation indices with vegetation fraction or cover [7,33,60], and in forest studies [8,61] report comparable figures. For example, Puliti et al. [8] achieved a 60% determination coefficient when estimating basal area from UAV imagery by combining point density and height. In our case, the canopy height model only improved the fit slightly (around 10%), probably due to the small size of most of the pines in both of our study sites. Later stages in the regeneration of vegetation would probably change the role of canopy height models in pine abundance prediction, as they would better allow discriminating pines from tall grasses and shrubs. Trials with canopy height of green pixels did not improve the fit of our models in this early stage (unpublished data).

Automatic exposure mode produced contrasting shutter speed values between sites, which translated into very diverse motion blur and radial distortion values. Surprisingly, we attained better results in La Carral, where motion blur was much larger than in Cal Rovira-Sanca. This larger motion blur resulted also in a marked radial distortion. Motion blur reduces the number of detected feature points and hence affects the quality of the photogrammetric processes [62]. A motion blur of two pixels has been deemed as enough to alter the results of these processes [63], although more recent work challenges this claim [62]. We estimated motion blur values below this threshold for the four flights, which could help explain the lack of relationship between motion blur and model fit performance at both sites. There were more evergreen shrubs in La Carral (eight shrubs) than in Cal Rovira-Sanca (one shrub) and the grid was built to avoid the green field at the eastern margin of Cal Rovira-Sanca area. Hence, we can rule out the presence of shrubs or other green vegetation as the reason for the lower performance of regressions in Cal Rovira-Sanca. The images in La Carral were taken with a much higher side and forward overlap, which could underlie the larger determination coefficients obtained in this site. Alternatively, the difference between flying times could have produced the difference between both sites. Lower sun angles result in changes in color temperature of the downwelling illumination [64,65], which would result in higher contrast between red and green channels.

We did not find marked differences between two contrasting flight altitudes in the general performance of greenness indices, although the selection of optimal greenness index or cell statistic depended on altitude at both sites. Flight height was an important determinant of the estimate of vegetation cover of wheat and barley experimental crops, together with growth stage and stitching software [33]. Flight height determines the final image resolution obtained as well as the effect of topography on radiance (by changing the relative angle between terrain slopes and the UAV) and hence, values of greenness indices [33,66]. A change of scale can be expected to modify the determination coefficient of greenness indices, by changing the greenness values of each pixel [67]. However, we did not notice any remarkable effect, probably because our analysis relied on descriptive statistics encapsulating information at a coarser spatial extent of 5*5 m (Figure S5). Topography was not expected to introduce important variations on the values of greenness indices, as it was relatively mild and homogeneous in our filed plots. Actually, in contrast with our expectations, at both study sites 120 m flights resulted in higher determination coefficients than 50 m ones. Given the differences among flights, in our case either higher overlap values or lower proportional motion blur could be the drivers of this trend. However, the obtained regression fits are much better in La Carral than in Cal Rovira-Sanca, despite their much higher motion blur. Hence, our results suggest that the key parameter determining the determination coefficient of sum of DBH on greenness indices is a higher image overlap, rather than motion blur.

The UAV deployment and imagery analysis we present here has its own limitations, leading to areas for further improvement of image acquisition and analysis. First, the camera used has a FOV of 110° and non-rectilinear lenses, which results in a heavy fish-eye effect and a significant barrel distortion. Although the processing of images within Agisoft Photoscan corrects for lens distortion before carrying out the image matching process, the effect of distortion on image quality is too high toward the edges of the image and may result in a low quality in important fractions of the orthoimage. Correcting the geometric distortion of images before loading them in the photogrammetric software could improve accuracy and geometry of the final orthomosaic. This correction can be carried out in specialized software and should also consider the distortion caused by the electronic rolling shutter and motion blur resulting from aircraft movement [23,62]. Newer UAVs harbor improved quality cameras, even at the low cost range of products, with rectilinear lenses and more limited geometric distortions, that would help improve the detection of photosynthetically active vegetation.

Second, trying to reduce complexity of deployment and analysis to a minimum, we did not place ground control points. Spatial accuracy of the obtained orthoimage was thus limited by the resolution of the PNOA imagery (pixel size of 25 cm and RMSE below 50 cm) [68] and by the error associated to the identification of common points in the obtained orthoimage and in the PNOA orthophotograph. Although most present day low cost UAVs geotag the images taken on the fly, their GPS accuracy is still low and commonly results in RMSEs above 1 meter [20]. Hence, if we aim to reduce costs to a minimum, we need to deploy several ground control points (GCP) around the plot to increase the spatial accuracy. The spatial accuracy we get when using GCPs will ultimately depend on the accuracy of GCP location measurement, which may be reduced to centimeters or even millimeters with differential (DGPS) or real time kinetic (RTK) GPS measurements [20]. A software-as-a-service (SaaS) can also improve accuracy at an affordable cost, allowing RMSE values around four times lower than those from built-in GPS devices of consumer-level UAVs [20].

Third, the camera we used records data only on the visual spectrum. While this is one of the strengths of our work (due to its availability and ease of operation), it also limits the capacity to properly discriminate the photosynthetically active vegetation from the dormant vegetation. Capturing near infrared radiation (NIR) requires a more expensive, multispectral sensor. These kind of multispectral cameras are now more readily available for UAVs deployment. The assessment of its value in forestry is under current development, but they will probably prove especially suitable for multitemporal image acquisition and comparison [13,69,70].

5. Conclusion

Our results show that low cost UAV platforms are a useful alternative to professional platforms when it comes to the detailed, cost-effective monitoring of forest ecosystems and forest recovery after disturbance, especially in non-profit forests where economic and human resources may be scarcer.

By using a simple approach with a consumer level UAV platform and low cost (educational license at 179$) or free software, we have predicted post-fire pine recovery (estimated as the sum of pine DBH) with a relatively high coefficient of determination of up to 60%. Furthermore, all the software used has been applied with its default values, avoiding tweaking of any of their parameters. This allows the use of these types of procedures by end users working in forestry and applied research, and at the same time leaves margin for further improving the accuracy of the process. All in all, consumer level UAVs can be expected to provide a common low cost tool for ecological monitoring of post-fire recovery and other conservation and monitoring tasks, given future drops in prices, increasing accuracy levels and widening in application types.

Supplementary Materials: The following are available online at http://www.mdpi.com/2504-446X/3/1/6/s1, Figure S1.1. A detail of a high-cover area within the Cal Rovira-Sanca area, Figure S1.2. A detail of a low-cover area within the La Carral area, Figure S2-1. General workflow followed for the analysis of UAV imagery and the software frameworks used in each of the steps. Numbers refer to the different intermediate outputs shown in figure S2, Figure S3-1. Height distribution of *Pinus nigra* and *Pinus sylvestris* in La Carral and Cal Rovira-Sanca, as measured in the field, Figure S3-2. Regressions of tree height on DBH for both species in La Carral and Cal Rovira-Sanca, Figure S4-1. Image overlap obtained for the four flights. Figures depict the number of images where each point is present, Figure S4-2. Image residuals for the Phantom Vision FC200 sensor after camera calibration performed within Agisoft PhotoScan software. Scale of residual lines is indicated in pixel units, Figure S5-1. Maps of the four greenness indices and the estimated canopy model from the four flights carried out at both sites and two flight altitude, Table S6-1. Estimates of the best fitting model for the 50 m high flight in La Carral, with one greenness index (ExGI) and the pine canopy height model (CHM), Table S6-2. Estimates of the best fitting model for the 120 m high flight in La Carral, with one greenness index (GCC) and the pine canopy height model (CHM), Table S6-3. Estimates of the best fitting model for the 50 m high flight in Cal Rovira-Sanca, with one greenness index (GCC) and the pine canopy height model (CHM), Table S6-4. Estimates of the best fitting model for the 50 m high flight in Cal Rovira-Sanca, with the greenness index (GCC) and one pine canopy height model (CHM), Figure S7-1. 3D scatterplots showing the overall best fitting model applied to each site and flight altitude.

Author Contributions: Conceptualization, A.R.L. and L.B.; methodology, A.R.L. and L.B.; software, A.R.L.; validation, A.R.L. and L.B.; formal analysis, A.R.L.; investigation, A.R.L. and L.B.; resources, L.B.; data curation, A.R.L.; writing—original draft preparation, A.R.L.; writing—review and editing, A.R.L. and L.B.; visualization, A.R.L.; supervision, L.B.; project administration, L.B.; funding acquisition, L.B.

Funding: ARL benefited from the support of the NEWFOREST Marie Curie IRSES project. Grant No. IRSES-GA-2013-612645.

Acknowledgments: We are indebted to Jan and Martí Brotons for their collaboration on field measurements.

Conflicts of Interest: The authors declare no conflict of interest.

References

1. Hardin, P.J.; Hardin, T.J. Small-scale remotely piloted vehicles in environmental research. *Geogr. Compass* **2010**, *4*, 1297–1311. [CrossRef]
2. Colomina, I.; Molina, P. Unmanned aerial systems for photgrammetry and remote sensing: A review. *ISPRS J. Photogramm. Remote Sens.* **2014**, *92*, 79–97. [CrossRef]
3. Matese, A.; Toscano, P.; Di Gennaro, S.F.; Genesio, L.; Vaccari, F.P.; Primicerio, J.; Belli, C.; Zaldei, A.; Bianconi, R.; Gioli, B. Intercomparison of UAV, aircraft and satellite remote sensing platforms for precision viticulture. *Remote Sens.* **2015**, *7*, 2971–2990. [CrossRef]
4. Wallace, L.; Lucieer, A.; Watson, C.; Turner, D. Development of a UAV-LiDAR system with application to forest inventory. *Remote Sens.* **2012**, *4*, 1519–1543. [CrossRef]
5. Getzin, S.; Nuske, R.S.; Wiegand, K. Using unmenned aerial vehicles (UAV) to quantify spatial gap patterns in forests. *Remote Sens.* **2014**, *6*, 6988–7004. [CrossRef]
6. Salamí, E.; Barrado, C.; Pastor, E. UAV fligh experiments applied to the remote sensing of vegetated areas. *Remote Sens.* **2014**, *6*, 11051–11081. [CrossRef]
7. Díaz-Varela, R.; de la Rosa, R.; León, L.; Zarco-Tejada, P.J. High-resolution airborne UAV imagery to assess olive tree crown parameters using 3D photo reconstruction: Application in breeding trials. *Remote Sens.* **2015**, *7*, 4213–4232. [CrossRef]
8. Puliti, S.; Ørka, H.O.; Gobakken, T.; Næsset, E. Inventory of small forest areas using and unmanned aerial system. *Remote Sens.* **2015**, *7*, 9632–9654. [CrossRef]

9. Guerra-Hernández, J.; González-Ferreiro, E.; Sarmento, A.; Silva, J.; Nunes, A.; Correia, A.C.; Fontes, L.; Tomé, M.; Díaz-Varela, R. Using high resolution UAV imagery to estimate tree variables in *Pinus pinea* plantation in Portugal. *For. Syst.* **2016**, *25*, 2171–9845. [CrossRef]

10. Karpina, M.; Jarząbek-Rychard, M.; Tymków, P.; Borkowski, A. UAV-based automatic tree growth measurement for biomass estimation. In Proceedings of the XXIII ISPRS Congress, Prague, Czech Republic, 12–19 July 2016; Volume XLI-G8. [CrossRef]

11. Baena, S.; Moat, J.; Whaley, O.; Boyd, D.S. Identifying species from the air: UAVs and the very high resolution challenge for plant conservation. *PLoS ONE* **2017**, *12*, e0188714. [CrossRef]

12. Thiel, C.; Schmullius, C. Comparison of UAV photograph-based and airborne lidar-based point clouds over forest from a forestry application perspective. *Int. J. Remote Sens.* **2016**, *38*, 2411–2426. [CrossRef]

13. Nebiker, S.; Annen, A.; Scherrer, M.; Oesch, D. A light-weight multispectral sensor for micro UAV-opportunities for very high resolution airborne remote sensing. *Int. Arch Photogram. Remote Sens. Spat. Inf. Sci.* **2008**, *XXXV*, 1193–1200.

14. Wallace, L.; Lucieer, A.; Malenobský, Z.; Turner, D.; Vopěnka, P. Assessment of forest structure using two UAV techniques: A comparison of airborne laser scanning and structure from motion (SfM) point clouds. *Forests* **2016**, *7*, 62. [CrossRef]

15. Gianetti, G.; Chirici, G.; Gobakkern, T.; Næsset, E.; Travaglini, D.; Puliti, S. A new approach with DTM-independent metrics for forest growing stock prediction using UAV photogrammetric data. *Remote Sens. Environ.* **2018**, *213*, 195–205. [CrossRef]

16. Puliti, S.; Talbot, B.; Astrup, R. Tree-stump detection, segmentation, classification, and measurement using unmanned aerial vehicle (UAV) imagery. *Forests* **2018**, *9*, 102. [CrossRef]

17. Blaschke, T. Object based image analysis for remote sensing. ISPRS J Photogram. *Remote Sens.* **2010**, *65*, 2–16. [CrossRef]

18. Chehata, N.; Orny, C.; Boukir, S.; Guyon, D. Object-based forest change detection using high resolution satellite images. In Proceedings of the Photogrammetric Image Analysis, Munich, Germany, 5–7 Octobre 2011; pp. 49–54.

19. Torres-Sánchez, J.; López-Granados, F.; Serrano, N.; Arquero, O.; Peña, J.M. High-trhoughput 3-D monitoring agricultural-tree plantations with unmanned aerial vehicle (UAV) technology. *PLoS ONE* **2015**, *10*, e0130479. [CrossRef]

20. Padró, J.C.; Muñoz, F.J.; Planas, J.; Pons, X. Comparison of four UAV georeferencing methods for environmental monitoring purposes focusing on the combined use with airborne and satellite remote sensing platforms. *Int. J. Appl. Earth Obs. Geoinf.* **2019**, *79*, 130–140. [CrossRef]

21. Rodrigo, A.; Retana, J.; Picó, X. Direct regeneration is not the only response of Mediterranean forests to large fires. *Ecology* **2004**, *85*, 716–729. [CrossRef]

22. Institut Cartogràfic i Geològic de Catalunya Mapa Geològic amb Llegenda Interactiva (Versió 2). 2014. Available online: http://betaportal.icgc.cat/wordpress/mapa-geologic-amb-llegenda-interactiva/ (accessed on 13 January 2017).

23. Vautherin, J.; Rutishauser, S.; Schneider-Zapp, K.; Choi, H.F.; Chovancova, V.; Glass, A.; Strecha, C. Photogrammetric accuracy and modelling of rolling shutter cameras. *ISPRS Ann. Photogramm. Remote Sens. Spat. Inf. Sci.* **2016**, *3*, 139–146. [CrossRef]

24. XnView MP 0.83 [GPL Software]. 2016. Available online: https://www.xnview.com/en/xnviewmp/ (accessed on 15 September 2016).

25. AgiSoft PhotoScan Professional (Version 1.2.6) [Software]. 2016. Available online: http://www.agisoft.com/downloads/installer/ (accessed on 13 September 2106).

26. Westoby, M.J.; Brasington, J.; Glasser, N.F.; Hambrey, M.J.; Reynolds, J.M. 'Structure-from-Motion' photgrammetry: A low-cost, effective tool for geoscience applications. *Geomorphology* **2012**, *179*, 300–314. [CrossRef]

27. *Agisoft PhotoScan User Manual. Professional Edition;* Version 1.4; Agisoft LLC: Saint-Petersburg, Russia, 2017.

28. McGaughey, R.J. *Fusion/LDV: Software for Lidar Data Analysis and Visualization. October 2016—Fusion Version 3.60;* US Department of Agriculture, Forest Service, Pacific Northwest Research Station: Seattle, WA, USA, 2016.

29. CloudCompare 2.7.0. [GPL software]. 2016. Available online: http://www.cloudcompare.org/ (accessed on 30 August 2016).

30. Kraus, K.; Pfeifer, N. Determination of terrain models in wooded areas with airborne laser scanner data. *ISPRS J. Photogramm. Remote Sens.* **1998**, *53*, 193–203. [CrossRef]
31. Woebbecke, D.M.; Meyer, G.E.; Von Bargen, K.; Mortensen, D.A. Color indices for weed identification under various soil, residue, and lighting conditions. *Trans. Am. Soc. Agric. Eng.* **1995**, *38*, 259–269. [CrossRef]
32. Sonnentag, O.; Hufkens, K.; Teshera-Sterne, C.; Young, A.M.; Friedl, M.; Braswell, B.H.; Milliman, T.; O'Keefe, J.; Richardson, A.D. Digital repeat photography for phenological research in forest ecosystems. *Agric. For. Meteorol.* **2012**, *152*, 159–177. [CrossRef]
33. Rasmussen, J.; Ntakos, G.; Nielsen, J.; Svensgaard, J.; Poulsen, R.N.; Christensen, S. Are vegetation indices derived from consumer-grade cameras mounted on UAVs sufficiently reliable for assessing experimental plots? *Eur. J. Agron.* **2017**, *74*, 75–92. [CrossRef]
34. Rouse, J.W.; Haas, R.H.; Schell, J.A.; Deering, D.W. Monitoring vegetation systems in the Great Plains with ERTS. In Proceedings of the 3rd ERTS Symposium, NASA SP-351 I, Washington, DC, USA, 10–14 December 1973; pp. 309–317.
35. Tucker, C.J. Red and photographic infrared linear combinations for monitoring vegetation. *Remote Sens. Environ.* **1979**, *8*, 127–150. [CrossRef]
36. Falkowski, M.J.; Gessler, P.E.; Morgan, P.; Hudak, A.T.; Smith, A.M.S. Characterizing and mapping forest fire fuels using ASTER imagery and gradient modeling. *For. Ecol. Manag.* **2005**, *217*, 129–146. [CrossRef]
37. Motohka, T.; Nasahara, K.N.; Oguma, H.; Tsuchida, S. Applicability of green-red vegetation index for remote sensing of vegetation phenology. *Remote Sens.* **2010**, *2*, 2369–2387. [CrossRef]
38. Viña, A.; Gitelson, A.A.; Nguy-Robertson, A.L.; Peng, Y. Comparison of different vegetation indices for the remote assessment of green leaf area index of crops. *Remote Sens. Environ.* **2011**, *115*, 3468–3478. [CrossRef]
39. Gitelson, A.A.; Kaufman, Y.J.; Stark, R.; Rundquist, D. Novel algorithms for remote estimation of vegetation fraction. *Remote Sens. Environ.* **2002**, *80*, 76–87. [CrossRef]
40. Leduc, M.-B.; Knudby, A.J. Mapping wild leek through the forest canopy using a UAV. *Remote Sens.* **2018**, *10*, 70. [CrossRef]
41. Toomey, M.; Friedl, M.A.; Frolking, S.; Hufkens, K.; Klosterman, S.; Sonnentag, O.; Balodcchi, D.D.; Bernacchi, C.J.; Biraud, S.C.; Bohrer, G.; et al. Greenness indices from digital cameras predict the timing and seasonal dynamics of canopy-scale photosynthesis. *Ecol. Appl.* **2015**, *25*, 99–115. [CrossRef] [PubMed]
42. QGIS Development Team. QGIS Geographic Information System. Open Source Geospatial Foundation Project. 2016. Available online: http://qgis.osgeo.org (accessed on 5 August 2016).
43. Ribbens, E.; Silander, J.A., Jr.; Pacala, S.W. Seedling recruitment in forests: Calibrating models to predict patterns of tree seedling dispersion. *Ecology* **1994**, *75*, 1794–1806. [CrossRef]
44. Parresol, B.R. Assessing tree and stand biomass: A review with examples and critical comparisons. *For. Sci.* **1999**, *45*, 573–593.
45. Bolte, A.; Rahmann, R.; Kuhr, M.; Pogoda, P.; Murach, D.; Gadow, K. Relationships between tree dimension and coarse root biomass in mixed stands of European beech (*Fagus sylvatica* L.) and Norway spruce (*Picea abies* [L.] Karst.). *Plant Soil* **2004**, *264*, 1–11. [CrossRef]
46. Poorter, L.; Bongers, L.; Bongers, F. Architecture of 54 moist-forest tree species: Traits, trade-offs, and functional groups. *Ecology* **2006**, *87*, 1289–1301. [CrossRef]
47. Poorter, L.; Wright, S.J.; Paz, H.; Ackerly, D.D.; Condit, R.; Ibarra-Manríquez, G.; Harms, K.E.; Licona, J.C.; Martínez-Ramos, M.; Mazer, S.J.; et al. Are functional traits good predictors of demographic rates? Evidence from five neotropical forests. *Ecology* **2008**, *89*, 1908–1920. [CrossRef]
48. He, H.; Zhang, C.; Zhao, X.; Fousseni, F.; Wang, J.; Dai, H.; Yang, S.; Zuo, Q. Allometric biomass equations for 12 tree species in coniferous and broadleaved mixed forests, Northeastern China. *PLoS ONE* **2018**, *13*, e0186226. [CrossRef]
49. Trasobares, A.; Pukkala, T.; Miina, J. Growth and yield model for uneven-aged mixtures of *Pinus sylvestris* L. and *Pinus nigra* Arn. In Catalonia, north-east Spain. *Ann. For. Sci.* **2004**, *61*, 9–24. [CrossRef]
50. Porté, A.; Bosc, A.; Champion, I.; Loustau, D. Estimating the foliage area of Maritime pine (*Pinus pinaster* Aït.) branches and crowns with application to modelling the foliage area distribution in the crown. *Ann. For. Sci.* **2000**, *57*, 73–86. [CrossRef]
51. Lefsky, M.A.; Harding, D.; Cohen, W.B.; Parker, G.; Shugart, H.H. Surface Lidar remote sensing of basal area and biomass in deciduous forests of Eastern Maryland, USA. *Remote Sens. Environ.* **1999**, *67*, 83–98. [CrossRef]

52. Cade, B.S. Comparison of tree basal area and canopy cover in habitat models: Subalpine forest. *J. Wildl. Manag.* **1997**, *61*, 326–335. [CrossRef]

53. Mitchell, J.E.; Popovich, S.J. Effectiveness of basal area for estimating canopy cover of ponderosa pine. *For. Ecol. Manag.* **1997**, *95*, 45–51. [CrossRef]

54. Popescu, S.C. Estimating biomass of individual pine trees using airborne lidar. *Biomass Bioenergy* **2007**, *31*, 646–655. [CrossRef]

55. R Core Team *R: A Language and Environment for Statistical Computing*; R Foundation for Statistical Computing: Vienna, Austria, 2016. Available online: https://www.R-project.org/ (accessed on 5 December 2016).

56. RStudio Team RStudio: Integrated Development for R. RStudio, Inc., Boston, MA, USA. 2015. Available online: http://www.rstudio.com/ (accessed on 5 December 2016).

57. Pearson, R.L.; Miller, L.D. Remote mapping of standing crop biomass for estimation of the productivity of the short-grass Prairie, Pawnee National Grassland, Colorado. In Proceedings of the 8th International Symposium on Remote sensing of Environment, Ann Arbor, MI, USA, 2–6 October 1972; pp. 1357–1381.

58. Baluja, J.; Diago, M.P.; Balda, P.; Zorer, R.; Meggio, F.; Morales, F.; Tardaguila, J. Assessment of vineyard water status variability by thermal and multispectral imagery using an unmanned aerial vehicle (UAV). *Irrig. Sci.* **2012**, *30*, 511–522. [CrossRef]

59. Bannari, A.; Morin, D.; Bonn, F.; Huete, A.R. A review of vegetation indices. *Remote Sens. Rev.* **1995**, *13*, 1–95. [CrossRef]

60. Geipel, J.; Link, J.; Claupein, W. Combined spectral and spatial modelling of corn yield based on aerial images and crop surface models acquired with an unmanned aircraft system. *Remote Sens.* **2014**, *11*, 10335–10355. [CrossRef]

61. Modzelewska, A.; Stereńczak, K.; Mierczyk, M.; Maciuk, S.; Balazy, R.; Zawila-Niedźwiecki, T. Sensitivity of vegetation indices in relation to parameters of Norway spruce stands. *Folia Forestalia Pol. Ser. A For.* **2017**, *59*, 85–98. [CrossRef]

62. Sieberth, T.; Wackrow, R.; Chandler, J.H. Motion blur disturbs—The influence of motion-blurred images in photogrammetry. *Photogramm. Rec.* **2014**, *29*, 434–453. [CrossRef]

63. Lelégard, L.; Delaygue, E.; Brédif, M.; Vallet, B. Detecting and correcting motion blur from images shot with channel-dependent exposure time. *ISPRS Ann. Photogramm. Remote Sens. Spat. Inf. Sci.* **2012**, *I-3*, 341–346.

64. Granzier, J.J.M.; Valsecchi, M. Variation in daylight as a contextual cue for estimating season, time of day, and weather conditions. *J. Vis.* **2014**, *14*, 1–23. [CrossRef] [PubMed]

65. Tagel, X. Study of Radiometric Variations in Unmanned Aerial Vehicle Remote Sensing Imagery for Vegetation Mapping. Master's Thesis, Lund University, Lund, Sweden, 2017.

66. Burkart, A.; Aasen, H.; Alonso, L.; Menz, G.; Bareth, G.; Rascher, U. Angular dependency of hyperspectral measurement over wheat characterized by a novel UAV based goniometer. *Remote Sens.* **2015**, *6*, 725–746. [CrossRef]

67. Mesas-Carrascosa, F.-J.; Torres-Sánchez, J.; Clavero-Rumbao, I.; García-Ferrer, A.; Peña, J.-M.; Borra-Serrano, I.; López-Granados, F. Assessing optimal flight parameters for generating accurate multispectral orthomosaics by UAV to support site-specific crop management. *Remote Sens.* **2015**, *7*, 12793–12814. [CrossRef]

68. Instituto Geográfico Nacional. PNOA—Características Generales. Available online: http://pnoa.ign.es/caracteristicas-tecnicas (accessed on 17 December 2018).

69. Del Pozo, S.; Rodríguez-Gonzálvez, P.; Hernández-López, D.; Felipe-García, B. Vicarious radiometric calibration of a multispectral camera on board an unmanned aerial system. *Remote Sens.* **2014**, *6*, 1918–1937. [CrossRef]

70. Suomalainen, J.; Anders, N.; Iqbal, S.; Roerink, G.; Franke, J.; Wenting, P.; Hünniger, D.; Bartholomeus, H.; Becker, R.; Kooistra, L. A lightweight hyperspectral mapping system and photogrammetric processing chain for unmanned aerial vehicles. *Remote Sens.* **2014**, *6*, 11013–11030. [CrossRef]

drones

MDPI

Article

Classification of Lowland Native Grassland Communities Using Hyperspectral Unmanned Aircraft System (UAS) Imagery in the Tasmanian Midlands

Bethany Melville [1,*], Arko Lucieer [2] and Jagannath Aryal [2]

[1] Faculty of Communication and Environment, Rhein-Waal University of Applied Sciences,
    47475 Kamp-Lintfort, Germany
[2] School of Technology, Environment and Design, College of Sciences and Engineering,
    Discipline of Geography and Environmental Sciences, University of Tasmania, Hobart 7001, Australia;
    Arko.Lucieer@utas.edu.au (A.L.); Jagannath.Aryal@utas.edu.au (J.A.)
*   Correspondence: bme@hsrw.eu

Received: 31 October 2018; Accepted: 23 December 2018; Published: 5 January 2019

Abstract: This paper presents the results of a study undertaken to classify lowland native grassland communities in the Tasmanian Midlands region. Data was collected using the 20 band hyperspectral snapshot PhotonFocus sensor mounted on an unmanned aerial vehicle. The spectral range of the sensor is 600 to 875 nm. Four vegetation classes were identified for analysis including *Themeda triandra* grassland, *Wilsonia rotundifolia*, *Danthonia/Poa* grassland, and *Acacia dealbata*. In addition to the hyperspectral UAS dataset, a Digital Surface Model (DSM) was derived using a structure-from-motion (SfM). Classification was undertaken using an object-based Random Forest (RF) classification model. Variable importance measures from the training model indicated that the DSM was the most significant variable. Key spectral variables included bands two (620.9 nm), four (651.1 nm), and 11 (763.2 nm) from the hyperspectral UAS imagery. Classification validation was performed using both the reference segments and the two transects. For the reference object validation, mean accuracies were between 70% and 72%. Classification accuracies based on the validation transects achieved a maximum overall classification accuracy of 93.

Keywords: hyperspectral; UAS; native grassland; random forest

---

1. Introduction

The Midlands region forms the primary agricultural region within the Australian State of Tasmania. The region was once populated by expanses of native grasslands and open woodlands [1]. However, these communities have seen a significant decline since European colonization began. Throughout subsequent years, native vegetation has been replaced by traditional European crop and forage species as agricultural land use in the region intensifies. Native vegetation communities still remain in the region, and are often used for grazing of sheep and cattle. However, the economic return associated with native grassland grazing is poorer than for introduced pasture species due to a lower nutritional value within the vegetation [2]. As a result, native grassland community extent has been steadily declining. Although the exact extent of native grassland vegetation lost is unknown, the estimated loss of community extent is estimated to be between 60% [3] and 90% [4] of a pre-colonial extent.

Collectively, the major grassland community types of the region are known as the lowland native grasslands. These communities form the Midlands biodiversity 'hotspot' [4], and contain an estimated 750 species, of which 85 are protected under Tasmanian or Federal Australian environmental protection laws [2,5]. The high level of biodiversity within these communities, coupled with the major threat of habitat loss due to expanding agricultural practices, has created a desperate need for novel approaches to mapping and monitoring of vegetation communities in the region. Community maps of native vegetation within the Midlands region are often incomplete or outdated [6], and, as a result, remote sensing has been proposed as a potential answer, due to its ability to provide frequently updateable maps of vegetation community extent and condition. However, due to the small patch size of remnant communities [4], coarse spatial resolution satellite-based approaches have proven to be moderately successful [7]. The rise of Unmanned Aerial Systems (UAS) in recent years, therefore, provides a unique opportunity to capture ultra-high spatial resolution data products that can be used to improve upon currently existing mapping approaches in the region.

The application of Unmanned Aircraft Systems (UAS) for environmental remote sensing applications has become increasingly prevalent in recent years. The ability of UAS to provide ultra-high spatial resolution datasets (<20 cm pixel size) at a relatively low cost makes them an attractive option for many researchers in this field [8]. The development of commercially available, 'off the shelf' platforms has led to a rapid increase in the applications for which UAS have been used in environmental sciences. The applicability of UAS for grassland monitoring and mapping is particularly attractive due to the ability of such systems to collect spatially detailed datasets on demand. This ability is integral to grassland remote sensing due to the high seasonal variability observed in communities [9–11].

Several studies have employed UAS as the principle platform in grassland research [12–14]. Although applications are primarily focussed on small-scale studies of agricultural productivity, such as estimating biomass [15], several studies have focussed on broader-scale ecological applications of UAS for various applications within grassland environments such as monitoring degradation and change [16], mapping species regeneration post-fire [17], estimating ground cover in rangelands [18], identifying grassland vegetation [19], and assessing species composition [13]. The most prevalent area of grassland research using UAS, however, is for rangeland monitoring and mapping. Extensive work has been undertaken, particularly in the South Western United States, to determine the feasibility of UAS for broad-scale, high spatial resolution analysis of semi-arid grassland and shrub communities [14,20,21].

The majority of remote sensing studies using UAS within the realm of ecological research have focused on the use of ultra-high spatial resolution datasets collected using broadband multispectral sensors [22] or RGB cameras [16] due to their low cost [8]. The use of broadband multispectral sensors is not always capable of providing sufficient spectral detail for accurate analysis of vegetation types and attributes, even when the data are acquired at high spatial resolutions. Applications of hyperspectral sensors using UAS platforms are still limited in general. However, there is an increasing body of work investigating their applicability in fields such as precision agriculture [23–28]. Due to the fact that the majority of previously available high spectral resolution sensors are based on push broom designs, the high fidelity Global Navigation Satellite System (GNSS) and Inertial Measurement Unit data were required for the creation of useable outputs [23,29]. This issue has led to limited use and application of UAS mounted hyperspectral sensors within the ecological remote sensing community. The development of frame-based and snapshot hyperspectral cameras, however, eliminates the need for complicated geometric processing, and makes the collection of hyperspectral datasets from UAS much more feasible. The use of such sensors has enormous potential for ecological vegetation mapping and monitoring due to the high degree of spectral information captured. This study aims to show the potential of a frame-based hyperspectral system for the vegetation community mapping in a highly heterogeneous grassland environment. Previous studies [7,30] in the area have identified a need to investigate the utility of hyperspectral systems in such environments, and this study aims to provide an important test case to improve lowland grassland community mapping through the use of novel sensor technologies.

2. Materials and Methods

2.1. Study Site and Vegetation Communities

In November 2015, imagery was collected at the Tunbridge Township Lagoon (42°08′52.36″, 147°25′45.50″), in the Tasmanian Midlands. The town of Tunbridge is located between the two major settlements of Hobart and Launceston, and marks the divide between the Northern and Southern Midlands regions. The lagoon serves as the only formally protected lowland native grassland habitat in Tasmania, and contains important remnant vegetation patches and many endangered species. The reserve covers an area of approximately 16 ha, and has wide floristic diversity. The western third of the site is populated by remnant *Themeda triandra* grassland and interspersed with *Acacia dealbata* and *Bursaria spinosa*. This portion of the site is steeply sloped in an easterly aspect. The remaining two thirds of the site are predominantly flat, and covered with a saltwater lagoon. The saltpan surrounding the lagoon is populated by many saline tolerant ground cover species, such as *Wilsonia rotundifolia*, *Sellieria radicans*, and, in places, the Australian Saltmarsh grass *Puccinellia stricta*. The areas between the saltpan and the bounding western and southern fences are populated by remnant *Danthonia trenuior* and *Poa labillardierie* grasslands. Vegetation communities are generally in good condition, although the southern side of the lagoon and a small area at the foot of the hill immediately adjacent to the lagoon is still recovering from unplanned burning in the summer of 2014.

For the purpose of this study, a subset of the total reserve area was targeted. This area is found on the south-western corner of the lagoon, and covers a transitional area between saltmarsh vegetation, native grassland communities dominated by *Danthonia trenuior* or *Poa labillardierei*, and the foot of the hill dominated by *Themeda triandra*. A total of four vegetation classes were identified for analysis, as well as a soil class. The first class consists of the saline vegetation communities found surrounding the lake including the succulent *Selliera radicans* and the ground cover *Wilsonia rotundifolia*. The second class covers the range of native grassland communities adjacent to the lagoon, which are called *Danthonia trenuior* and *Poa labillardierie* dominated areas. The common feature among these communities is that they all follow the C_3 photosynthetic pathway. The third class covers the *Themeda triandra* remnant patches found on the western slopes of the site. The fourth class is representative of the scattered *Acacia* and *Bursaria* specimens found among the *Themeda* grassland, and the final class consists of exposed soils found within the lagoon.

2.2. Data Collection

Data was collected using a multi-rotor UAS (DJI S1000) for hyperspectral imagery, and a fixed-wing UAS (Phantom FX-61) for RGB imagery. Hyperspectral imagery was collected using a PhotonFocus MV1-D2048x1088-HS02-96-G2-10 (www.photonfocus.com), which is a 25 band hyperspectral snapshot camera, with a spectral wavelength range from 600 to 875 nm and average FWHM (Full-width Half Maximum) of 6 nm. The wavelength range of the camera was selected based on previous research identifying this spectral region as containing key areas of separability for lowland native grassland communities [7,30]. The camera houses a hyperspectral chip manufactured by IMEC with 25 band-pass filters mounted on top of the sensor's pixels in a 5 × 5 mosaic pattern. The 25 bands are captured simultaneously and the pixels are organised in a hypercube of 409 by 216 pixels, and resampled to 20 bands after spectral correction. Table 1 gives the central wavelength for each of the 20 bands. The camera captured images at 4 frames per second (fps). We used a 16 mm focal length lens providing a field of view of 39° and 21° horizontal and vertical, respectively. The camera was mounted on a gimbal on a DJI S1000 multi-rotor UAS, and flown in a grid survey pattern at 80 m above ground level with a flight line separation of 22 m providing 60% side overlap between flight strips and 97% forward overlap. The ground sampling distance (GSD) of the raw imagery was 3 cm, but, after spatial and spectral resampling, this was reduced to 15 cm. The flight track was recorded with a navigation-grade global navigation satellite system (GNSS) receiver (zti communications Z050 timing and navigation module, spatial accuracy 5–10 m), and each hyperspectral image frame was geotagged based on GPS time. One hundred images were captured before the flight with the lens cap on the camera and averaged to collect a dark current image. Another 100 images were captured on the ground of a Spectralon panel directly before and after UAS flights to apply a vignetting lens correction and to allow for conversion of DN values to reflectance. A Python script was developed to process the raw camera data into hypercubes with reflectance values. The resulting images were exported to the GeoTiff format and imported into AgiSoft Photoscan (with their corresponding GPS coordinates). The Structure from Motion (SfM), dense matching, model generation, and orthophoto generation processing steps were performed in a Photoscan based on band 14 (801 nm). Additionally, 22 photogrammetric ground control points were randomly distributed across the study site and coordinated with a dual frequency geodetic-grade RTK GNSS receiver (Leica 1200), which resulted in an absolute accuracy of 2 to 4 cm. A 348 m by 255 m hyperspectral ortho-mosaic of the full scene was produced for further analysis. Sky conditions were clear and sunny during all UAS flights. The hyperspectral flights occurred during a one-hour time window around solar noon. Figure 1 shows an overview of the study site using the RGB UAV imagery, as well as the footprint of the hyperspectral dataset. Figure 2 shows the hyperspectral orthophoto loaded as a false-color RGB composite using bands 14, 5, and 1 (801.7 nm, 668.1 nm, and 612.8 nm). Reflectance values are shown for the subset areas of each class.

Table 1. Spectral band designations for the 20-band hyperspectral PhotonFocus dataset.

Band Number	1	2	3	4	5	6	7	8	9	10	11	12	13	14	15	16	17	18	19	20
Center Wavelength (nm)	612	620	643	651	668	676	684	712	737	751	763	776	789	801	813	825	842	854	864	872

Figure 1. Overview of Tunbridge Township Lagoon showing extent of the lake, and distribution of vegetation communities. The extent flown by the hyperspectral sensor within the large site is shown in red.

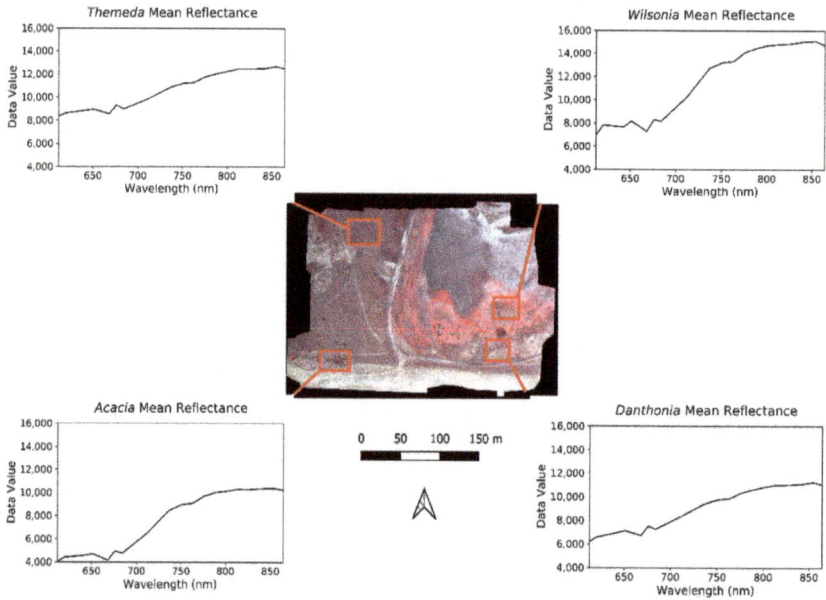

Figure 2. Vegetation communities found within the study site, as represented by the 14 cm ortho-mosaic. Central panel shows the overview of the study site, with image subsets showing general appearance of classes within the scene. Example spectral signatures are given for each class, based on the 20 spectral bands of the orthomosaic, as collected by the PhotonFocus hyperspectral sensor.

Additionally, an RGB camera was flown on a fixed-wing UAS at a height of 80 m (Sony alpha 5100, 20 mm focal length lens, FOV = 60° × 43°, shutter speed 1/1000 s, GSD = 1.7 cm, forward overlap 80%, side overlap: 70%). An RGB orthophoto mosaic was produced 1.7 cm spatial resolution in Agisoft Photoscan using the SfM workflow described earlier. The Ground Control Points (GCPs) were used in the bundle adjustment, which produced an RGB orthophoto mosaic with an absolute spatial accuracy of 2 cm. A 15 cm spatial resolution digital surface model (DSM) was derived from the 3D dense point cloud and triangulated model. From this DSM, the slope was derived using the surface toolset in ArcGIS 10.3 [31]. The RGB DSM and Slope model were used in the analysis over the hyperspectral outputs due to a better horizontal and vertical accuracy.

For validation purposes, two 100 m transects, as shown in Figure 3, were established at the site during aerial data acquisition. The transects covered the *Wilsonia*, *Danthonia*, and *Themeda* classes over an area in which the communities intergrade significantly. Transects were run east to west across the center of the study area. Observations of plant communities were taken every meter along each transect. A polygon representing the observation area was then digitized in ArcGIS 10.3 for each point along the transect, and assigned the relevant class based on the field observations. Training points for the classification model were based on field observations acquired from 5 × 5 meter transects established in November and December 2015. Transect centroids were generated based on random stratification within the site and the coordinates of ground control points used for the UAS data acquisition. For each transect, a tape measure was aligned north to south, and the vegetation community was recorded every 1.25 m along the tape for a total of five observations. Observations were then taken east to west every 2.5 m, for nine observations. The majority vegetation community for the transect area was then determined and used as the classification label. For each transect centroid, a GPS coordinate was acquired, and imported into ArcGIS 10.3. Based on each point, a five meter buffer was generated and a series of points spaced 15 cm apart were generated within the bounds of the buffer zone. Furthermore, 15 cm was selected as the point spacing since this matched the spatial resolution of the final ortho-mosaic generated from the UAS dataset. Vegetation classes were assumed to be uniform within the entire 5 m zone, and care was taken to ensure that no transitional transects were used for training purposes.

Lastly, a set of reference segments representing homogeneous class regions were digitized for the four vegetation classes to serve as additional validation. Object size varied relative to class extent, which was between 1 m² and 100 m². Additional validation data was manually digitized in order to create an adequate sample size for classification validation and because there were no recorded observations for the *Acacia* class, based on the 100 m transects. It was found that the two 100 m transects failed to provide an appropriate number of validation points for some vegetation classes. Figure 3 shows the distribution of the 5 m buffered training zones, the validation transects, and manually digitized reference polygons within the study site.

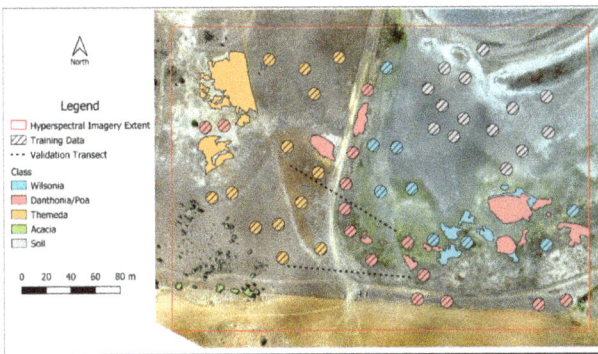

Figure 3. Distribution of training and validation points throughout the study site. Hashed areas indicate data used for training, while un-hashed polygons were used for validation.

2.3. Random Forest Training and Classification

Image segmentation was performed using the Multiresolution Segmentation Algorithm [32] in eCognition based on the 20 spectral bands of the orthomosaic, the DSM, and the slope model. A scale factor of 1700 was used, with the compaction factor set to 0.1, and the shape factor set to 0.9. The DSM and slope model were included in the classification approach since they were previously found to be of high importance for classification of these communities [7]. The DSM was not converted to a canopy height model due to the low height of vegetation in some sites of the study area (<1 cm in some areas). Once the segmentation had been completed, a random forest model was trained for classification. Training and classification were performed on the 20 band ortho-mosaic, the DSM, and the slope layers. Since the number of input variables was equal to 22, the number of variables to try (mtry) was set equal to 4, since the established optimal parameter value is equal to $\sqrt{m}$, where m is the number of variables used [33,34]. Internal cross-validation accuracies were obtained for the model, in addition to variable importance measures. Validation of the classification results was performed twice including once using the digitised reference objects, and a second time based on the 100 m transects. The reference objects and transects were not merged into a single validation dataset due to the difference in the scale of analysis between the two datasets. Merging the transect observation areas into the larger reference segment area would, therefore, result in this fine spatial scale of the observations being lost due to the large discrepancy in the area of analysis between the two datasets. Validation was performed using the reference segments in order to provide a large-scale estimate of accuracy across the entire scene, and also to ensure that an adequate number of validation points was used for each class. Validation using homogeneous reference segments also enables the evaluation of misclassification due to over-segmentation. The field transects were used as a secondary source of validation since they provide valuable data about the sensitivity of the sensor and classification results to transitional zones between communities. The high spatial frequency of observations along the transects allows for accurate determination of the exact point of change between vegetation types. Since community intergrading has been identified previously as being a significant source of classification confusion for these communities [7], the decision was made to collect data capable of evaluating the sensitivity of the segmentation scale and classification approaches.

3. Results

3.1. RF Training and Variable Importance Measures

Table 2 shows the confusion matrix and training accuracies obtained from the RF internal cross-validation. Class values are given as a pixel count, while accuracy is given as a percentage. The overall training accuracy was 97.44%. The obtained accuracies are high for all classes, with the *Themeda* and *Acacia* classes having slightly lower accuracies than the *Wilsonia* and *Danthonia/Poa* classes. There is very little confusion between classes, which indicates good potential class separability within the dataset.

Table 2. RF training accuracy and confusion matrix for all classes. Confusion matrix values are given as a pixel count, while accuracy is reported as a percentage.

	Wilsonia	*Danthonia*	*Themeda*	*Soil*	*Acacia*	Accuracy (%)
Wilsonia	35,565	277	0	36	0	99.13
Danthonia	343	54,725	693	0	1	98.14
Themeda	0	3473	48,293	0	19	93.26
Soil	3	0	0	55,778	0	99.99
Acacia	0	123	345	0	7740	94.29

Figure 4 shows the variable importance measures obtained from the RF training model for each class. The DSM has a very high importance score relative to the other variables, and is identified as highly important for all classes. The most important spectral bands are bands two (620.9 nm), seven (684.9 nm), and eleven (763.2 nm). The *Danthonia* class has high importance values for these bands compared to the other classes. The *Themeda* class has the highest importance value for the DSM.

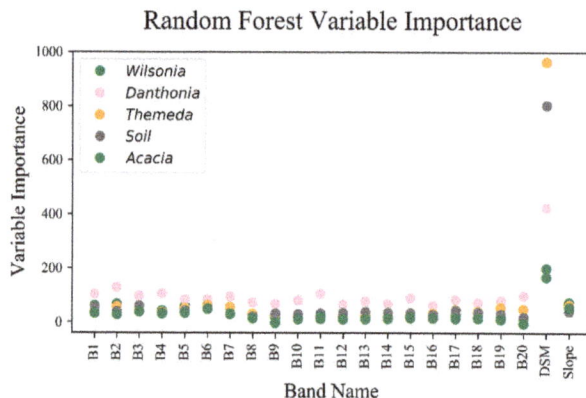

Figure 4. RF variable importance models obtained from the training model for each class.

3.2. RF Classification Results and Accuracy

Figure 5 shows the classification results obtained from the RF model. Delineation of class boundaries is clear and matches expected distributions. There is some misclassification between

Wilsonia and the soil class evident in the center of the scene. The transitional zones between the two grassland types *Danthonia/Poa* and *Themeda* are clearly demarked. Areas of disturbance within the *Themeda* community are also clearly visible. There is significant omission within the *Acacia* class.

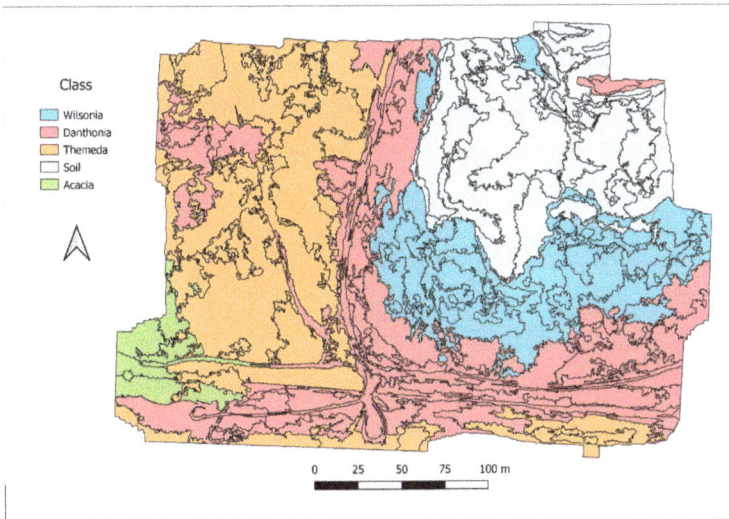

Figure 5. Random Forest classification results for all community classes. General class delineation is good. However, there is a significant omission and commission of the Acacia class, and some observable confusion between the Wilsonia and Danthonia classes.

Table 3 gives the final RF confusion matrix and User's and Producer's accuracies based on the accuracy assessment undertaking using the digitized reference segments. The final overall accuracy for the result is 71.8%. The Users' accuracy was obtained by calculating the percentage of all image objects identified as a given class that were correctly identified. Producers' accuracy was obtained by calculating how much of the reference area for a given class was correctly classified. Users' accuracies for the *Wilsonia, Themeda,* and *Acacia* classes are 98%, 92.4%, and 92.6%, respectively. Additionally, the users' accuracies for the *Wilsonia* and *Themeda* classes show similar values. The producer's accuracy for the *Acacia* class was also low at 41.9%. The producer's accuracy for the *Danthonia* class is 98.4%.

Table 3. Confusion matrix, User's Accuracy, and Producer's Accuracy for all classes based on evaluation against the manually digitised reference segments. All values are percentages.

	Wilsonia	*Danthonia*	*Themeda*	*Acacia*	**Soil**	**User's**	**Producer's**
Wilsonia	98	1.9	0	0	0.01	98.0	38.4
Danthonia	54.5	41.3	4.2	0	0	41.3	98.4
Themeda	0	0	94.2	5.7	0	92.4	95.3
Acacia	0	0	7.4	92.6	0	92.6	41.9

Table 4 reports the validation results based on the two 100 m transects. A mean User's accuracy of 85.1% was achieved. The type of confusion is similar to that observed in the evaluation based on the reference segments. Primarily, confusion occurs between the *Wilsonia* and *Danthonia* classes. General patterns of classification accuracies for the three classes were similar to those in the previous assessment, with *Wilsonia* having poorer performance than the two grassland classes. Since there were no recorded observations of the *Acacia* class along either of the two 100 m transects, the class has been excluded from the accuracy assessment here.

Table 4. Confusion matrix and per class User's and Producers' accuracy for the three vegetation classes covered by the validation transects. Confusion matrix values are given as percentages.

	Wilsonia	*Danthonia*	*Themeda*	User's	Producer's
Wilsonia	100	0	0	100	73.2
Danthonia	13.9	86.1	0	86	83.1
Themeda	0	23.4	76.6	76.2	100

4. Discussion

The accuracies obtained for the RF training model are much higher than the accuracies obtained for any of the final classification results. The presence of significant discrepancies between validation and training accuracies can indicate potential bias in the sampling regime, or unrepresentative training datasets. Since the number of input points was high (~250,000), it was decided that a single RF model was to be derived. The high spatial resolution of the dataset (14 cm) means that very fine-scale variations in species composition can potentially be detected. Since training points were derived over a 5 m plot area to emulate the conditions, there is potential to incorporate multiple thematic classes within a single plot. Conversely, however, the use of small numbers of widely dispersed pixels may lead to a failure to properly account for class variability in the training and classification stages of analysis.

The variable importance measures obtained show a very high importance for the two topographic variables. Spectral variables have significantly lower importance values for each class, and all classes have their highest importance level recorded for the DSM. The high importance value assigned to the DSM in the *Themeda* class is due to the class occurring exclusively on the hilly area within the study site, whereas the *Danthonia* and *Wilsonia* classes occur on the flat saltpan surrounding the lagoon. This difference is of key importance due to similarities in the canopy structure between the *Themeda* and *Danthonia* classes, which may potentially lead to confusion between the two classes during classification.

Classes exhibit different importance levels across the range of spectral input bands, with the *Danthonia* class having the highest overall spectral importance values. Key bands of importance for this class include band two (620.9 nm), 11 (763.2 nm), and 14 (801.7 nm). The observed regions of importance align with those identified in previous research [7]. All other vegetation classes have comparatively low spectral importance values. Low importance values for the spectral variables is likely a result of high correlation between the bands. The high values for spectral bands in the *Danthonia* result is likely due to similarities between the physical distribution of the *Danthonia* and *Wilsonia* classes, since both only occur on the same flat region of the saltpan. Since these two classes intergrade significantly, and cannot be differentiated based on topography, spectral bands are the only available source for differentiation with the RF model. This is likely to contribute to the poor performance of the class overall since previous studies undertaken at this site indicate great difficulty discerning between the *Danthonia* and *Wilsonia* classes due to their similar photosynthetic pathway and phenological staging [30].

The primary source of inaccuracy in the classification results was confusion between the *Danthonia* and *Wilsonia* classes. The confusion in this case was only in one direction, in that a large proportion of the *Danthonia* class was erroneously classified as *Wilsonia*, while there was very little misclassification of *Wilsonia* as *Danthonia*. As the two communities intergrade extensively, the establishment of discrete reference objects for validation was very difficult even within the two 100 meter transects. Inaccuracy when creating these reference objects is likely to be the case of the poor overall accuracy obtained for the *Danthonia* class. Since the two classes occur in the same geographic area, primarily on low-lying saltpan, the DSM and slope variables are not likely to increase separability between the two classes.

The *Danthonia* class exhibits significantly different final classification accuracies between the validation performed using the reference objects and the validation based on the image transects. The observed differences in classification accuracies between assessments based on the validation transects and manually digitized objects indicate that there is a need to collect high spatial resolution validation datasets in order to accurately assess the performance of classifications in this case.

This paper shows that discrimination between similar vegetation communities can be successful using hyperspectral, frame-based UAS mounted sensors. The spectral range of the sensor used in this case is quite narrow, with bands only covering the red and near-infrared portions of the spectrum from 600 to 875 nm. The use of a sensor with an expanded range into the visible portion of the spectrum may potentially improve classification outcomes, due to an increased ability to detect unique spectral properties of communities. The ability of the sensor to successfully discriminate between communities is most apparent in the results of the two grassland communities. The main difference between these communities is that they have different photosynthetic pathways. Many studies have shown that differentiation of grassland species based on photosynthetic pathway and phenological variation is an appropriate method [35,36]. The dataset used in this scenario was acquired at a time of year where the *Themeda* grassland was entering a period of senescence, and the *Danthonia/Poa* grassland was beginning its annual growth cycle. The results shown in this paper reiterate the findings of previous studies that have shown that phenology is a critical component of the grassland community identification. The overall findings of this paper suggest that frame-based hyperspectral UAS mounted sensors can be used to successfully differentiate between native grassland communities with a high degree of accuracy.

5. Conclusions

This paper presents the results of an RF classification approach for identifying lowland native grassland communities in the Tasmanian Midlands using high spatial and spectral resolution UAS imagery. The findings of this study indicate that high spectral resolution UAS datasets can provide detailed community discrimination at a fine spatial scale, and show great potential for community and species level mapping. The higher classification accuracies obtained for the transect validations indicate that accurate assessment of community gradients requires the collection of high spatial frequency field observations over a large area. The small extent covered by the two transects means that assessment of communities in this manner is limited for this result. Future studies could benefit from the creation of multiple transects with closely spaced observations to aid in a more robust assessment of classification results. This paper presents an important case-study that show-cases the use of a hyperspectral UAS mounted sensors for ecological community classification. The results of this study indicate that the sensor used here is particularly useful in the discrimination of grassland communities.

Author Contributions: B.M., A.L., and J.A. designed the experiment. B.M. and A.L. performed the field work and collected the UAS imagery. A.L. created the ortho-mosaic and digital surface model. Melville performed the image analysis and analyzed the results. B.M. and A.L. prepared the manuscript, with J.A. providing additional editorial device.

Funding: This research was funded by a Tasmanian Postdoctoral Research Award granted by the University of Tasmania.

Acknowledgments: The authors would like to acknowledge the support of Stephen Harwin and Darren Turner in the collection of field datasets. Additionally, the authors would like to thank Louise Gilfedder and Jamie Kirkpatrick for their advice and guidance.

Conflicts of Interest: The authors declare no conflict of interest.

References

1. Fensham, R.; Kirkpatrick, J. The conservation of original vegetation remnants in the midlands, tasmania. *Pap. Proc. R. Soc. Tasman.* **1989**, *123*, 229–246.

2. Mokany, K.; Friend, D.; Kirkpatrick, J.; Gilfedder, L. *Managing Tasmanian Native Pastures: A Technical Guide for Graziers*; Tasmanian Institute of Agricultural Research: Hobart, Australia, 2006.
3. Gilfedder, L. Threatened Species From Tasmania's Remnant Grasslands. *Tasforests* **1990**, *129–132*.
4. Beeton, R. *Advice to the Minister for the Environment, Heritage and the Arts from the Threatened Species Scientific Committee (the Committee) on an Amendment to the List of Threatened Ecological Communities under the Environment Protection and Biodiversity Conservation*; DEWHA: Hobart, Australia, 2006.
5. DEWHA. *Lowland Native Grasslands of Tasmania*; DEWHA: Hobart, Tasmania, 2010.
6. Kirkpatrick, J.B.; Gilfedder, L.A.; Fensham, R.J. *City Parks and Cemeteries: Tasmania's Remnant Grasslands and Grassy Woodlands*; Tasmanian Conservation Trust: Hobart, Tasmania, 1988.
7. Melville, B.; Lucieer, A.; Aryal, J. Object-based random forest classification of Landsat ETM+ and WorldView-2 satellite imagery for mapping lowland native grassland communities in Tasmania, Australia. *Int. J. Appl. Earth Obs. Geoinf.* **2018**, *66*, 46–55. [CrossRef]
8. Anderson, K.; Gaston, K.J.; Anderson, K.; Gaston, K.J. Lightweight unmanned aerial vehicles will revolutionize spatial ecology. *Front. Ecol. Environ.* **2013**, *11*, 138–146. [CrossRef]
9. Goetz, S. Multi-sensor analysis of NDVI, surface temperature and biophysical variables at a mixed grassland site. *Int. J. Remote Sens.* **1997**, *18*, 71–94. [CrossRef]
10. Wen, Q.; Zhang, Z.; Liu, S.; Wang, X.; Wang, C. Classification of Grassland Types by MODIS Time-Series Images in Tibet, China. *IEEE J. Sel. Top. Appl. Earth Obs. Remote Sens.* **2010**, *3*, 404–409. [CrossRef]
11. Xu, B.; Yang, X.C.; Tao, W.G.; Miao, J.M.; Yang, Z.; Liu, H.Q.; Jin, Y.X.; Zhu, X.H.; Qin, Z.H.; Lv, H.Y.; et al. MODIS-based remote-sensing monitoring of the spatiotemporal patterns of China's grassland vegetation growth. *Int. J. Remote Sens.* **2013**, *34*. [CrossRef]
12. Laliberte, A.S.; Rango, A.; Herrick, J. Unmanned aerial vehicles for rangeland mapping and monitoring: A comparison of two systems. In Proceedings of the Americal Society of Photogrammetry and Remote Sensing Annual Conference, Tampa, FL, USA, 7–11 May 2007.
13. Lu, B.; He, Y.; Liu, H. Investigating species composition in a temperate grassland using Unmanned Aerial Vehicle-acquired imagery. In Proceedings of the Fourth International Workshop on Earth Observation and Remote Sensing Applications, Guangzhou, China, 4–6 July 2016.
14. Rango, A.; Laliberte, A.; Herrick, J.; Winters, C.; Havstad, K.; Steele, C.; Browning, D. Unmanned aerial vehicle-based remote sensing for rangeland assessment, monitoring and management. *J. Appl. Remote Sens.* **2009**, *3*, 033542.
15. Kawamura, K.; Sakuno, Y.; Tanaka, Y.; Lee, H.; Lim, J.; Kurokawa, Y.; Watanabe, N. Mapping herbage biomass and nitrogen status in an Italian ryegrass (*Lolium multiflorum* L.) field using a digital video camera with balloon system. *J. Appl. Remote Sens.* **2011**, *5*, 053562. [CrossRef]
16. Svoray, T.; Perevolotsky, A.; Atkinson, P.M. Ecological sustainability in rangelands: The contribution of remote sensing. *Int. J. Remote Sens.* **2013**, *34*, 6216–6242. [CrossRef]
17. Silva, B.; Lehner, L.; Roos, K.; Fries, A.; Rollenbeck, R.; Beck, E.; Bendix, J. Mapping Two Competing Grassland Species from a Low-Altitue Helium Balloon. *IEEE J. Sel. Top. Appl. Earth Obs. Remote Sens.* **2014**, *7*, 3038–3049. [CrossRef]
18. Breckenridge, R.P.; Dakins, M.; Bunting, S.; Harbour, J.L.; Lee, R.D.; Breckenridge, R.P.; Dakins, M.; Bunting, S.; Harbour, J.L.; Lee, R.D. Using Unmanned Helicopters to Assess Vegetation Cover in Sagebrush Steppe Ecosystems. *Rangel. Ecol. Manag.* **2016**, *65*, 362–370. [CrossRef]
19. Burai, P.; Tomor, T.; Beko, L.; Deak, B. Airborne hyperspectral remote sensing for identification grassland vegetation. In Proceedings of the International Archives of Photogrammetry, Remote Sensing and Spatial Information Sciences, La Grande Motte, France, 28 September–3 October 2015; Volume XL, pp. 427–431.
20. Laliberte, A.S.; Herrick, J.E.; Rango, A.; Winters, C. Acquisition, Or thorectification, and Object-based Classification of Unmanned Aerial Vehicle (UA V) Imager y for Rangeland Monitoring. *Photogramm. Eng. Remote Sens.* **2010**, *76*, 661–672. [CrossRef]
21. Laliberte, A.S.; Goforth, M.A.; Steele, C.M.; Rango, A. Multispectral Remote Sensing from Unmanned Aircraft: Image Processing Workflows and Applications for Rangeland. *Remote Sens.* **2011**, *3*, 2529–2551. [CrossRef]
22. Gao, J. Quantification of grassland properties: How it can benefit from geoinformatic technologies? *Int. J. Remote Sens.* **2006**, *27*, 1351–1365. [CrossRef]

23. Aasen, H.; Burkart, A.; Bolten, A.; Bareth, G. Generating 3D hyperspectral information with lightweight UAV snapshot cameras for vegetation monitoring: From camera calibration to quality assurance. *ISPRS J. Photogramm. Remote Sens.* **2015**, *108*, 245–259. [CrossRef]

24. Berni, J.A.J.; Zarco-tejada, P.J.; Suárez, L.; Fereres, E. Thermal and Narrowband Multispectral Remote Sensing for Vegetation Monitoring From an Unmanned Aerial Vehicle. *IEEE Trans. Geosci. Remote Sens.* **2009**, *47*, 722–738. [CrossRef]

25. Gevaert, C.M.; Suomalainen, J.; Tang, J.; Kooistra, L. Generation of Spectral—Temporal Response Surfaces by Combining Multispectral Satellite and Hyperspectral UAV Imagery for Precision Agriculture Applications. *IEEE J. Sel. Top. Appl. Earth Obs. Remote Sens.* **2015**, *8*, 3140–3146. [CrossRef]

26. Zarco-Tejada, P.J.; Gonzalez-Dugo, V.; Berni, J.A.J. Remote Sensing of Environment Fluorescence, temperature and narrow-band indices acquired from a UAV platform for water stress detection using a micro-hyperspectral imager and a thermal camera. *Remote Sens. Environ.* **2012**, *117*, 322–337. [CrossRef]

27. Zarco-Tejada, P.J.; Guillén-climent, M.L.; Hernández-clemente, R.; Catalina, A. Agricultural and Forest Meteorology Estimating leaf carotenoid content in vineyards using high resolution hyperspectral imagery acquired from an unmanned aerial vehicle (UAV). *Agric. For. Meteorol.* **2013**, *171–172*, 281–294. [CrossRef]

28. Zarco-Tejada, P.J.; Ustin, S.L.; ML, W. Temporal and Spatial Relationships between Within-Field Yield Variability in Cotton and High Spatial Hyperspectral Remote Sensing Imagery. *Agron. J.* **2005**, *97*, 641–653. [CrossRef]

29. Aasen, H.; Bendig, J.; Bolten, A.; Bennertz, S.; Willkomm, G.; Bareth, M. Introduction and preliminary results of a calibration for full-frame hyperspectral cameras to monitor agricultural crops with UAVs. *Int. Arch. Photogramm. Remote Sens. Spat. Inf. Sci.* **2014**, *XL-7*, 5194. [CrossRef]

30. Melville, B.; Lucieer, A.; Aryal, J. Assessing the Impact of Spectral Resolution on Classification of Lowland Native Grassland Communities Based on Field Spectroscopy in Tasmania, Australia. *Remote Sens.* **2018**, *10*. [CrossRef]

31. *ESRI ArcGIS Desktop: Release 10.3 2014*; Environmental Systems Research Institute: Redlands, CA, USA.

32. Baatz, M. Schäpe, a Multiresolution Segmentation: An optimization approach for high quality multi-scale image segmentation. In *Angewandte Geographische Informationsverarbeitung XII. Beiträge zum AGIT-Symposium Salzburg*; Herbert Wichmann Verlag: Karlsruhe, Germany, 2000; pp. 12–23.

33. Breiman, L. Random Forests. *Mach. Learn.* **2001**, *45*, 5–32. [CrossRef]

34. Naidoo, L.; Cho, M.A.; Mathieu, R.; Asner, G. Classification of savanna tree species, in the Greater Kruger National Park region, by integrating hyperspectral and LiDAR data in a Random Forest data mining environment. *ISPRS J. Photogramm. Remote Sens.* **2012**, *69*, 167–179. [CrossRef]

35. Dube, T.; Mutanga, O.; Elhadi, A.; Ismail, R. Intra-and-inter species biomass prediction in a plantation forest: Testing the utility of high spatial resolution spaceborne multispectral rapideye sensor and advanced machine learning algorithms. *Sensors* **2014**, *14*, 15348–15370. [CrossRef] [PubMed]

36. Foody, G.M.; Dash, J. Estimating the relative abundance of C3 and C4 grasses in the Great Plains from multi-temporal MTCI data: Issues of compositing period and spatial generalizability. *Int. J. Remote Sens.* **2010**, *31*, 351–362. [CrossRef]

drones

MDPI

Article

Rapid Assessment of Ecological Integrity for LTER Wetland Sites by Using UAV Multispectral Mapping

Ricardo Díaz-Delgado [1],*, Constantin Cazacu [2] and Mihai Adamescu [2]

[1] Remote Sensing and GIS Laboratory (LAST-EBD), Estación Biologica de Doñana, CSIC, Avda. Américo Vespucio 26, 41092 Sevilla, Spain

[2] Research Center for Systems Ecology and Sustainability, University of Bucharest, Spl. Independentei 91–95, 050095 Bucharest, Romania; constantin.cazacu@g.unibuc.ro (C.C.); adacri@gmail.com (M.A.)

* Correspondence: rdiaz@ebd.csic.es; Tel.: +34-954-232340

Received: 12 November 2018; Accepted: 19 December 2018; Published: 23 December 2018

Abstract: Long-term ecological research (LTER) sites need a periodic assessment of the state of their ecosystems and services in order to monitor trends and prevent irreversible changes. The ecological integrity (EI) framework opens the door to evaluate any ecosystem in a comparable way, by measuring indicators on ecosystem structure and processes. Such an approach also allows to gauge the sustainability of conservation management actions in the case of protected areas. Remote sensing (RS), provided by satellite, airborne, or drone-borne sensors becomes a very synoptic and valuable tool to quickly map isolated and inaccessible areas such as wetlands. However, few RS practical indicators have been proposed to relate to EI indicators for wetlands. In this work, we suggest several RS wetlands indicators to be used for EI assessment in wetlands and specially to be applied with unmanned aerial vehicles (UAVs). We also assess the applicability of multispectral images captured by UAVs over two long-term socio-ecological research (LTSER) wetland sites to provide detailed mapping of inundation levels, water turbidity and depth as well as aquatic plant cover. We followed an empirical approach to find linear relationships between UAVs spectral reflectance and the RS indicators over the Doñana LTSER platform in SW Spain. The method assessment was carried out using ground-truth data collected in transects. The resulting empirical models were implemented for Doñana marshes and can be applied for the Braila LTSER platform in Romania. The resulting maps are a very valuable input to assess habitat diversity, wetlands dynamics, and ecosystem productivity as frequently as desired by managers or scientists. Finally, we also examined the feasibility to upscale the information obtained from the collected ground-truth data to satellite images from Sentinel-2 MSI using segments from the UAV multispectral orthomosaic. We found a close multispectral relationship between Parrot Sequoia and Sentinel-2 bands which made it possible to extend ground-truth to map inundation in satellite images.

Keywords: UAVs; ecological integrity; LTER; LTSER; multispectral mapping; ground-truth; Parrot Sequoia; Sentinel-2

1. Introduction

Rapid assessment of ecosystem status, both functioning and structure, has become a major requirement for managers and conservationists in choosing response to disturbances or understanding global change effects at local scale [1]. Humanity is failing to make sufficient progress in confronting grand environmental challenges, and alarmingly, most of them are getting far worse [2]. One of the most informative ways to retrieve a quick conservation status picture of either habitats or species is based on survey and long-term monitoring programs [3,4]. Long-term ecological research (LTER) networks are informing about the factors driving changes in biodiversity, the self-organizing capacity of ecosystems, the effects of rare events and disturbances, the impacts of stressors on ecosystem function, and the interactions between short- and long-term trends [5]. These LTER networks rely on site-based monitoring and research by providing data and detecting trends identifying drivers and pressures. However, there is a need for methods and parameters harmonization in order to enhance sites comparisons and identify global patterns. The regional European LTER network, named LTER-Europe (www.lter-europe.net), has developed and adopted a new framework to easily derive ecosystems state from selected indicators: the ecological integrity (EI) framework. The idea of EI is based on the principle for precaution against ecological risks in the framework of sustainable development. The EI framework combines biotic and abiotic aspects of ecosystems with ecosystem structures and processes [6,7]. Its aim is to safeguard relevant ecosystem services and preserve the capability to continue self-organized development of systems and services, by becoming more complex systems or adapting to change. The ecological integrity framework has enabled indicator selection and enhanced data integration and upscaling for individual LTER sites. In addition, in 2007, the LTER-Europe network introduced the concept of long-term socio-ecological research (LTSER). This approach extends LTER concept to coupled socio-ecological (or human-environment) systems. LTSER aims to provide a knowledge base that helps to reorient socioeconomic trajectories towards more sustainable pathways [8]. LTSER platforms, inside LTER-Europe network (Figure 1a), are extensive landscapes characterized by manifold interactions between society and nature, ranging from strict conservation areas to intensively used ones [9].

While seeking fast assessment of large and inaccessible areas such as LTSER wetlands, synoptic tools become essential to provide the required integrative view. For this purpose, managers turn usually either to the use of in situ measurements or estimates from automatic or handheld sensors and probes or to ad-hoc sampling procedures [10]. In such cases, data collected by these means can point out local changes or trends but it will seldom inform on spatial gradients or reveal under-sampled locations. At this point, remote sensing imagery becomes the major contributor to spatially visualize and locate any kind of environmental threat or disturbance such as wildfires, eutrophication processes, flooding, etc. Optical images captured from Earth observation mid-resolution satellites (tens of meters) are widely available for free, such as the ones captured by Landsat or Sentinel missions. However, high- and very-high-resolution images (from centimeters to few meters) are costly and have to be pre-ordered and programmed to be acquired over the study area. In the former case, scenes are periodically acquired enabling to build a time series of images to address temporal changes and trends at the landscape scale. In the latter, finer resolution allows for detailed habitat mapping for instance, while dramatically increasing costs. The same is also true for airborne photogrammetric campaigns either with photogrammetrical, multispectral, or hyperspectral sensors on board of planes. Conversely, UAVs can be flown over the same area as frequently as required, only constrained by weather conditions or legislation, becoming a suitable monitoring tool for any target. As a major trait, UAVs provide the opportunity to define spatial resolution as detailed as requested according to the mission objectives [11] constraining the total area covered per unit of time. Yet it is not the only role played by UAVs as they can play the role of ground-truth for other sensors either airborne or onboard of satellites, while overflying large and remote or inaccessible areas. Rapid growth of commercial UAVs and affordable prices together with the increase on the innovative offer of miniaturized multispectral sensors and cameras is widely spreading their use across the globe. Just a few years ago, dealing with mission planning and post-processing was also a challenge. Nowadays, integrated solutions such as the one provided by SenseFly© are offering accurate and endurable platforms such as the Sensefly eBee able to carry different airborne cameras. Thermal, visible or multispectral cameras can be mounted alone in its single pod. The company also provides a very integrative post-processing solution with Pix4D© software, enabling to carry out complete missions in very short time for many applications on protected areas. Actually, UAVs are mainly used to monitor crop health and status by upscaling physiological variables applying model inversion methods [12]. Very few studies have addressed similar topic for other environments such as rangelands but none for natural wetlands [13,14].

In this paper, we assess the applicability of UAV borne multispectral cameras for fast mapping of the ecological state of two LTER wetlands following the ecological integrity framework. We evaluate several indices proposed to retrieve the necessary information to assess the inundation level, the plant and open water cover, plant height, and water turbidity and depth. We confirm that the derived maps can contribute to enhance and enlarge the area to be used as ground-truth data for satellite remote sensing images (in this case, we used Sentinel-2 images). The easiness and high performance of multispectral cameras on board of fixed wing UAV is demonstrated while offering fast EI assessment and ground-truth for satellite remote sensing images.

2. Study Sites and Conservation Issues

2.1. The Doñana LTSER Platform

The Doñana LTSER Platform is located SW of Spain (Figure 1c). It is a UNESCO Biosphere Reserve, a Ramsar Site, and a Natural World Heritage Site. It includes the largest wetland in Western Europe and a large dune ecosystem with its respective shoreline and representative terrestrial plant communities. The area is home to many species, including the Iberian lynx and the imperial eagle. The Doñana marshes play a critical role as a stopover, breeding and wintering point for thousands of European, Iberian and African birds. The long-term ecological monitoring program focuses on threatened species and habitats and uses a multi-scale approach [15]. Conservation objectives include the preservation of critically endangered species, the abundance of waterfowl, and the protection of Mediterranean wetlands and terrestrial ecosystems. Data are systematically collected on vegetation, threatened flora, limnology, mammals, birds, amphibians, and reptiles in an integrative way [4]. Doñana marshes, which cover an extent of 260 km^2, provide important ecosystem services such as aesthetic, spiritual, scientific, and eco-tourism provided by waterbirds under the cultural domain or grazing for cattle under the provisioning domain and nutrient cycling and water purification as regulating or supporting services. [16,17]. So, mapping inundation levels, hydroperiod, water turbidity and depth, together with aquatic plant cover becomes essential to characterize 'within-habitat structure', 'habitat cover', or 'water quality'.

2.2. The Braila Island LTSER Platform

The Braila Island LTSER platform is located in the small island of Braila in the Danube River, southeast Romania (Figure 1b). The Small Island of Braila is especially rich in bird species. Together with the coastal Danube Delta, the wetland system is an important stepping stone for bird migration routes in southeastern Europe. This socio-ecological system is inhabited by near 300,000 people and comprises heavily modified ecosystems (e.g., Big Island of Braila) but also systems under a natural functional regime (e.g., Small Island of Braila), being of a crucial natural and socio-economical value. Most of the area has been drained for agricultural purposes. As a consequence, connectivity between the Danube and the floodplains is very limited [18]. The Danube river in the Braila Islands section has been ranked as a heavily modified water body according to criteria 2.1 (embankment works) due to the hydro-technical works on 79% of the river stretch sector and a candidate to "heavily modified" according with the WFD criteria 2.2 (regulation works) as a result of dredging of 21% of the river bed for intensive navigation. The main remnant of the natural floodplains consists in the wetlands from the Small Island of Braila Natural Park with a total surface of 210 km^2 and the floodplains between the riverbanks and dikes of almost 93 km^2 [19]. Water quality in this stretch of the Danube River is also rated as moderate. The main pollution sources are agriculture, industry, navigation, and domestic households. Also in Braila Island, periodic mapping of inundation levels, water turbidity and aquatic plant cover as well as floodplain tree cover are crucial to identify sudden changes in management affecting flooding.

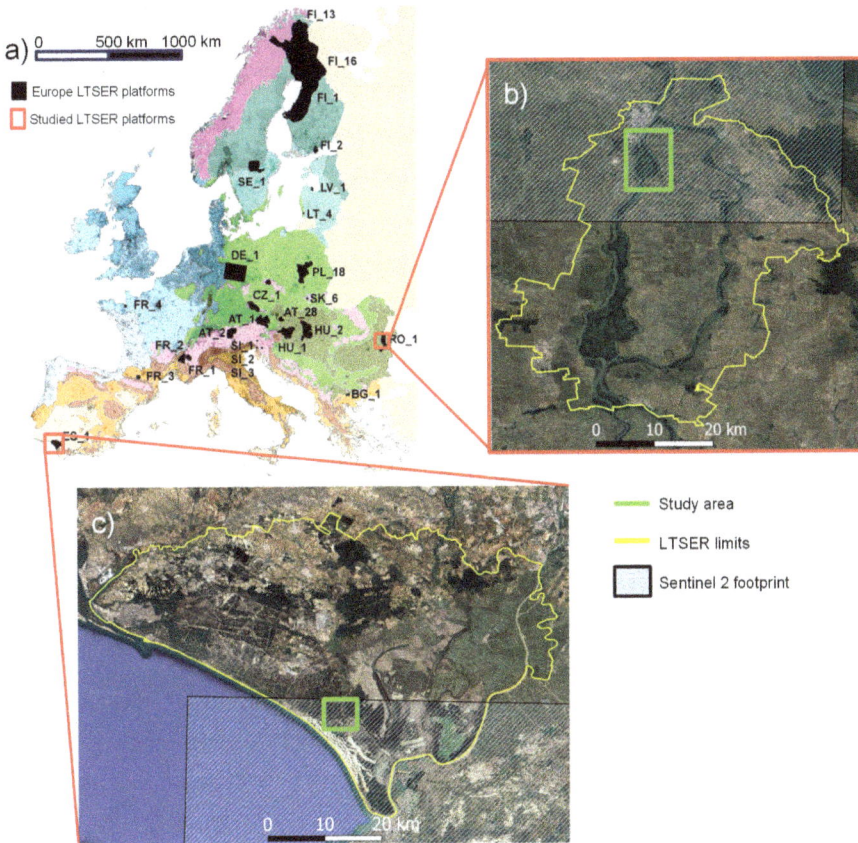

Figure 1. (**a**) Location of the 28 LTSER platforms in Europe on top of biogeographical regions (modified from Mirtl et al. [9]). Many more have been created in other LTER regional networks [20]. (**b**) Zoom in at Braila Island LTSER platform limits (yellow line) and the study area where we carried out the UAV flight (green square). (**c**) Zoom in at Doñana LTSER platform limits (yellow line) and the study area (red line).

3. Materials and Methods

3.1. Practical Remote Sensing Indicators for Rapid Ecological Integrity (EI) Assessment of Wetlands

Structural EI components are based on biotic diversity and abiotic heterogeneity. The components of processes (input, output, storage) are related to energy, matter and water balances. Structural and process components are interrelated and may be used to reflect states, changes and pressures enabling fast assessment of the protected area. However, although the recent work by Haase et al. [5] conveys the links between EI indicators and the essential biodiversity variables (EBV) [21], there is still much work to do in order to retrieve such valuable information using remote sensing tools [22,23].

In our case, we are dealing with rapid assessment of wetlands sharing similar pressures and drivers. As water is the main agent defining states, the most significant indicator is water presence or inundation level. Water presence or absence in the wetlands is essentially informative to water input at the water budget component of processes in the EI framework (Table 1). While revisiting the same place, hydroperiod can be easily retrieved as a function of inundation residence through time, informing on water storage for the wetland. Table 1 shows some examples of remote sensing indicators, which can be easily mapped and directly related to EI indicators in wetlands (rivers, lakes or coastal shallow wetlands). Some references are also provided, although mostly based on satellite remote sensing we include many of them using UAVs borne sensors. Many more references on every indicator are available in the literature.

Table 1. Examples and references of remote sensing applications related to EI indicators in wetlands. Those in red were used in this study. Modified from Haase et al. [5].

	Elements of Ecological Integrity	Indicators of Ecological integrity	Examples for Remote Sensing Indicators	References
Structures	Biotic Diversity	Flora Diversity	Aquatic plant cover mapping (emergent, floating, submerged)	[24–26]
			Floodplain forest species mapping	[27,28]
			Alien species mapping	[29–32]
		Fauna Diversity	Productivity estimates in birds colonies	[33,34]
			Animals abundance estimates with thermal mapping	[35–37]
			Input for Species Distribution	[38–40]
		Within Habitat Structure	Aquatic plant height	[41,42]
			Land use mapping in catchment	[43]
			Landscape indicators (connectivity, fragmentation)	[44]
	Abiotic Heterogeneity	Water	Water turbidity	[45,46]
			Water delineation, water depth	[45,47,48]
			Water temperature	[49]
		Atmosphere	Water vapour content	[50]
			Net radiation	[51]
		Habitat	Digital terrain models	[52,53]
Processes	Energy Budget	Input	Fraction absorbed of Photoshynthetic Active Radiation (FaPAR)	[54,55]
		Storage	Chlorophyll concentration in open water bodies	[56,57]
			Net Primary Production	[58,59]
		Output	Phenology	[60]
			Albedo	[61]
			Heat Flux (SEB models)	[62]
	Matter Budget	Input	Water colour as a proxy for nutrients availability	[63]
			Algal blooms	[64]
			Sedimentation processes	
		Storage	Aquatic plants biomass	[16,65]
		Output	Mapping of Grazing intensity	
	Water Budget	Input	Inundation mapping	[66]
		Storage	Hydroperiod	[67]
			Water level estimated from water depth	
		Output	Evapotranspiration	[68]

3.2. Multispectral Camera, UAV Mission Planning, and Image Processing

Among the wide offer of drones and cameras, we selected the eBee solution consisting of a Parrot Sequoia multispectral camera integrated in the eBee fixed wing UAV [69]. The choice was based both on the spectral bands (b1 green -550@40nm-, b2 red -660@40nm-, b3 red edge -735@10nm- and b4 near infrared -790@40nm-) provided by Parrot Sequoia and the large flying extent provided by the eBee plane (up to 40 ha in one single flight of 25', with flight height 120 m and pixel size 11 cm). Additionally, sequoia camera brings a sensor of irradiance in the upper side of the sensor which is concurrently capturing irradiance while taking pictures [70] and a RGB sensor of higher resolution. A calibration panel is provided with every unit to be pictured before flight allowing for bands reflectance calculation after flight [71].

Missions over Doñana marshes in Doñana LTSER platform and Braila LTSER platform were designed to cover an inundation gradient (Figure 2). Both sites were flown with clear sky conditions around 12 UTC at the maximum legal altitude (120 m above the terrain), perpendicular to the dominant wind direction and beyond-visual-line-of-sight (BVLOS). In the case of Doñana marshes, the flight covered the ecotone area between the sandy substrates and the marsh with variable inundation levels and aquatic plant cover (Figure 2a). In order to guarantee safe and dry operation we took-off and landed on the sandy substrates of the surrounding area. As the Small Island of Braila can only be accessed by boat and is covered by dense floodplain forests, flight mission was designed according to such constraints. Therefore, take-off and landing operations were carried out from the opposite river bank (Figure 2b). Doñana flight was accomplished on 22 April 2017 and lasted for 25 minutes. Braila flight took 27 minutes and was acquired on 1 August 2017.

Figure 2. UAV flight missions at (**a**) Doñana marshes and (**b**) Small Island of Braila. Image courtesy of 2018 Google©.

Radiometric calibration was simply achieved by reflectance calculation according to radiance coefficients and irradiance measured at every picture center [72]. Pictures are geotagged with the SenseFly eMotion software using the UAV flight logs and the set of pictures were introduced into Pix4D© software to be stitched and generate a multispectral orthomosaic together with digital surface model [73]. Figure 3 shows the general workflow of the study. Ground control points (GCP) for geometric correction could not be established because inside the marshes we could not find any valid lineal or conspicuous element to be used as reference in the flight area and we realized that the use of artificial targets (made of canvas fabric) once placed over water were easily displaced by wind. Visual geometric validation was carried out with the available high resolution images for every site (Google satellite or Bing satellite) with recognizable features such as tree/shrub canopies, paths and fences at the edge of the marsh. Absolute root-mean-square error (RMSE) was calculated using seven visually recognizable points in both layers.

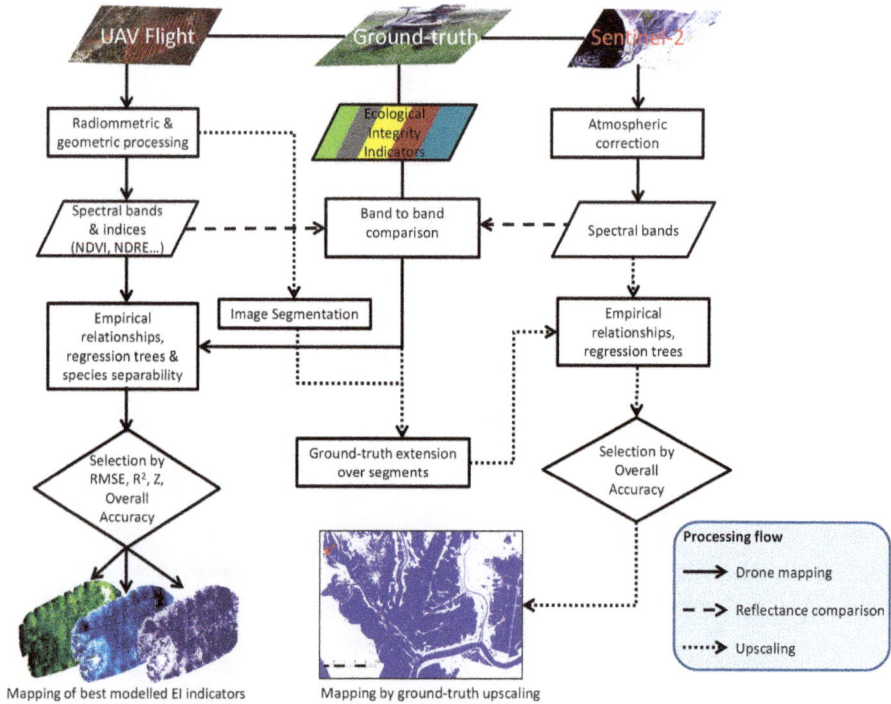

Figure 3. Methodological workflow followed in this study.

3.3. Ground-Truth Sampling, Accuracy Assessment, and Remote Sensing Wetland Indicators Mapping

In order to assess UAV mapping accuracy we carried out field sampling over the Doñana marsh study area immediately after the flight and Sentinel-2 image acquisition of the day. Ground-truth was collected by walking the marshes following predesigned regular transects to maximize the total area sampled across different inundated areas (Figure 4). Sampling points were located every 60 m in visually homogeneous sites at least for a radius of 15 m. Different wetland indicators were collected at every sampling point being originally representative of the 30 × 30 m Landsat TM and ETM+ pixel size according to Díaz-Delgado et al. [67] methodology being also valid for Sentinel-2 10 × 10 m pixel size. We recorded data contributing to EI indicators such as water turbidity, water depth, percentage of bare ground, aquatic plant and open water cover, and plant species abundance and dominance (Table 2). Geolocation of every point was recorded by means of PDA-GPS units with less than 3.3 m horizontal position error on average.

Table 2. RS wetland indicators collected as ground-truth with their categories and the spectral bands and indices used in this study.

RS Wetland Indicator	Categories/Range	Spectral Bands/Indices
Water turbidity	Continuous (1.41–471)	Water turbidity index (WTI [46]) Normalized difference water index (NDWI [47])
Water depth	Continuous (0–57)	Normalized difference red edge (NDRE [74])
Plant cover per plant type (emergent)	0%, 1–5%, 5–25%, 25–75%, >75%	Normalized difference vegetation index (NDVI [75]) Normalized difference red edge (NDRE)
Plant height	Continuous (3-150)	Green band, vegetation height model (VHM)
Percentage of open water	0%, 1–5%, 5–25%, 25–75%, >75%	NIR band
Inundation	Non-inundated and Inundated	NIR band
Dry bare-ground cover	0%, 1–5%, 5–25%, 25–75%, >75%	Not tested
Plant type	Emergent, floating, submerged, algae	Not tested

Field data was used to analyze statistical relationships between Sequoia spectral bands and different spectral indices (Table 2) applicable to retrieve wetlands EI indicators.

A random selection of 70% of ground data were used to explore linear modeling and a set of 30% to independently test accuracy of every model. The assessment was based on the values of coefficient of determination, R^2 and RMSE. Accordingly, we used the best lineal fit to map the tested wetland indicators. For inundation mapping we applied regression tree technique to discriminate between inundated and non-inundated classes based on Sequoia NIR band reflectance values [67]. In this case, classification accuracy was assessed with overall agreement (OA) and Kappa index. For plant height we also explored the relationship with vegetation height model (VHM) obtained from the subtraction of digital surface model (DSM) and digital terrain model (DTM) as applied by Bendig et al. [76]. DSM and DTM are generated from the point cloud by Pix4D software using the photograms of the RGB camera [77].

Finally, we assessed the discriminative ability of Sequoia multispectral bands to separate spectral signatures of the most common dominant aquatic vegetation species including emergent, floating, and submerged plants. We performed a separability analysis to assess the best band to discriminate among the different pairs of aquatic plant species. For this purpose, we used the normalized distance Z [78] which provides high values for the most different compared species.

3.4. Reflectance Comparison with Satellite Images

Both drone missions were set to be coincident with Sentinel-2 (S2) MSI acquisitions over the study sites. Thus, S2 images were available for the same dates such as 22 April for Doñana and 1 August for Small Island of Braila. S2 images were downloaded by using the semi-automatic classification (SAC) plug-in implemented in QGIS [79]. An atmospheric correction is carried out in the pre-processing of the S2 images based on the basic dark-object-subtract (DOS) technique, inspired in Chavez [80] and Moran et al. [81]. DOS method approximates the path radiance value of a given band from the minimum value of the histogram (dark object), assuming an intrinsic reflectance of the darkest object (1%). The rest of the radiance received by the satellite sensor proceeds from the atmospheric path, and must be then subtracted from every pixel before dividing the at-sensor spectral radiance by the irradiance. The model assumes that the transmittance is 1. The SAC-QGIS method is quick and simple, fully image-based, avoiding the need for atmospheric auxiliary data to perform the correction. It does not account for topographic effects, a characteristic that is not relevant in the present study due to the flat morphology of the study area.

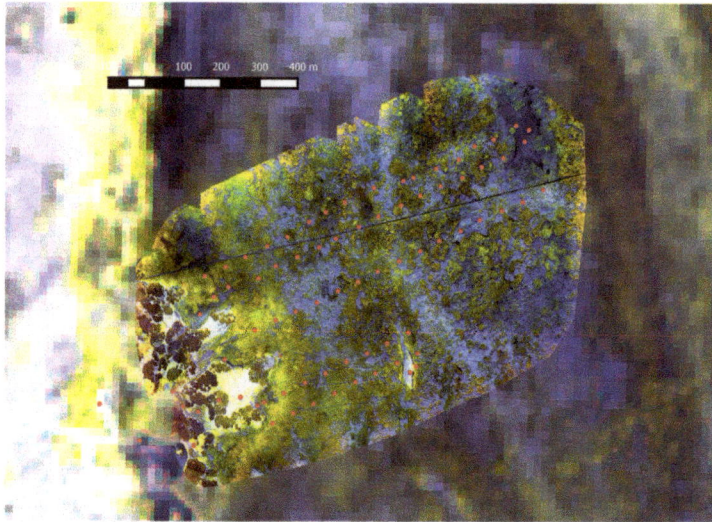

Figure 4. Location of ground-truth sampling points (red dots) in the Doñana study area. Background is composed by Sequoia multispectral orthomosaic on top of Sentinel-2 image equivalent wavelength composites (4-3-2 and 7-6-4).

Ground-truth points were used to compare Doñana Sequoia and S2 multispectral bands (Figure 3). The S2 bands selected for comparison with Sequoia bands were the most similar in terms of spectral resolution: B3 (560@45 nm), B4 (665@38 nm), B6 (740@18 nm), and B7 (783@28 nm). Bands 6 and 7 with original pixel size of 20 m were resampled to the B3 and B4 spatial resolution (10 m). Assessment was carried out by comparing R^2 and RMSE values for band reflectance relationships (Figure 3). Reflectance for Sequoia bands was obtained calculating the average of the pixels contained in one S2 pixel.

3.5. Upscaling of Ground-Truth Data

An upscaling essay was carried out to extend the information collected from ground-truth data to the whole S2 image. We applied a segmentation on Sequoia multispectral orthomosaic. Sequoia multispectral image was segmented using the four spectral bands using 'segment mean shift' available in ArcGIS with the following specs: spectral detail 15.5, spatial detail 15, minimum segment size 20 pixels (2.60 m). Then we made a spatial assignation of ground point data to the whole extension of the resulting spatially and spectrally homogeneous segments. Therefore, we used the labeled segments to build regression tree for inundation mapping but this time using S2 spectral data from the pixels inside the segments. The optimal threshold value was used to map inundation for the full S2 scene. Overall agreement and kappa index were used as indicators for accuracy assessment.

4. Results

4.1. Geometric Accuracy of UAV Multispectral Orthomosaics

Table 3 shows the geometric characteristics of every mission. Despite the fact we could not set up GCPs, the absolute root-mean-square errors (RMSE) of the orthomosaics were below 40 cm. Such values are still useful to assess the RS wetlands indicators as a function of ground-truth data collected to be homogeneous in 15 m around the point [67].

Figure 5 shows a visual assessment of geometric accuracy for the Doñana flight. Several distance measurements between both layers in recognizable objects revealed geometric matching below 0.40 m.

Table 3. Geometric characteristics of the UAV missions carried out over Doñana and Braila.

Flight Characteristics	Doñana	Braila
Ground sampling distance (cm)	12.85	14.03
Area covered (ha)	88.75	91.71
Number of images	1712	1380
Lateral overlap (%)	60	60
Longitudinal overlap (%)	80	80
Absolute RMS error (cm)	34	35.5

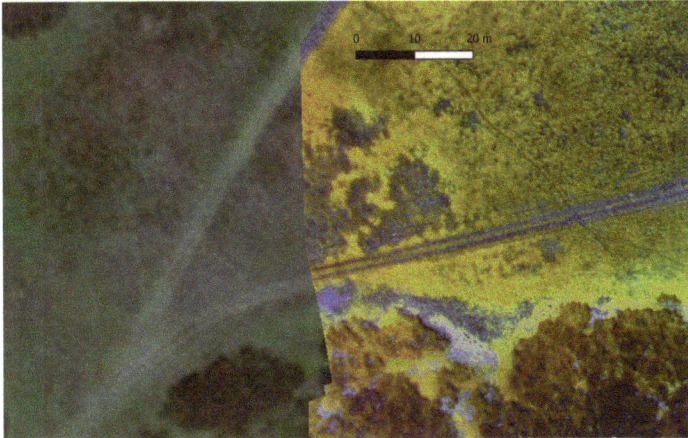

Figure 5. Overlay showing the edge of Sequoia multispectral orthomosaic on top of Bing Satellite in an area where two tracks and plant canopies were used to visually check geometric accuracy.

The eBee plane travelled a total distance of 15.5 km over the Doñana study area and 12.5 km across the Small Island of Braila.

4.2. Spectral Modeling of Remote Sensing Wetlands Indicators

A total of 73 ground-truth points out of 75 collected in the field, were finally used to analyze linear relationships between Parrot Sequoia bands and different spectral indices with the wetlands RS indicators. Figure 6 shows some of the resulting maps by implementing the most significant relationships found for several RS indicators.

While Sequoia NDRE (Table 2) showed a significant and positive linear relationship with percent cover of emergent aquatic plants (R^2 = 0.67; Figure 6a), plant height of these helophytes showed a significant but very low negative correlation with the Sequoia green band. Vegetation height from the subtraction of digital surface and terrain models did not show a significant relationship (R^2 = 0.01, p > 0.1). The best predictor for percentage of open water was found to be the NIR band (Sequoia band 4) showing a significant linear relationship (R^2 = 0.46, Figure 6b) as expected according to the extinction of this wavelength in water bodies (Figure 6). However, water depth variation was significantly explained by NDRE but in a very weak manner (R^2 = 0.36; RMSE = 13 cm) so we did not applied it.

Water turbidity variability was explored with visible bands and indices showing a significant and positive relationship only for open water areas (>50% open water) with NDWI better than the one with WTI. Accordingly, we applied the mapping model by using the classes >50% with percentage of open water (Figure 6c).

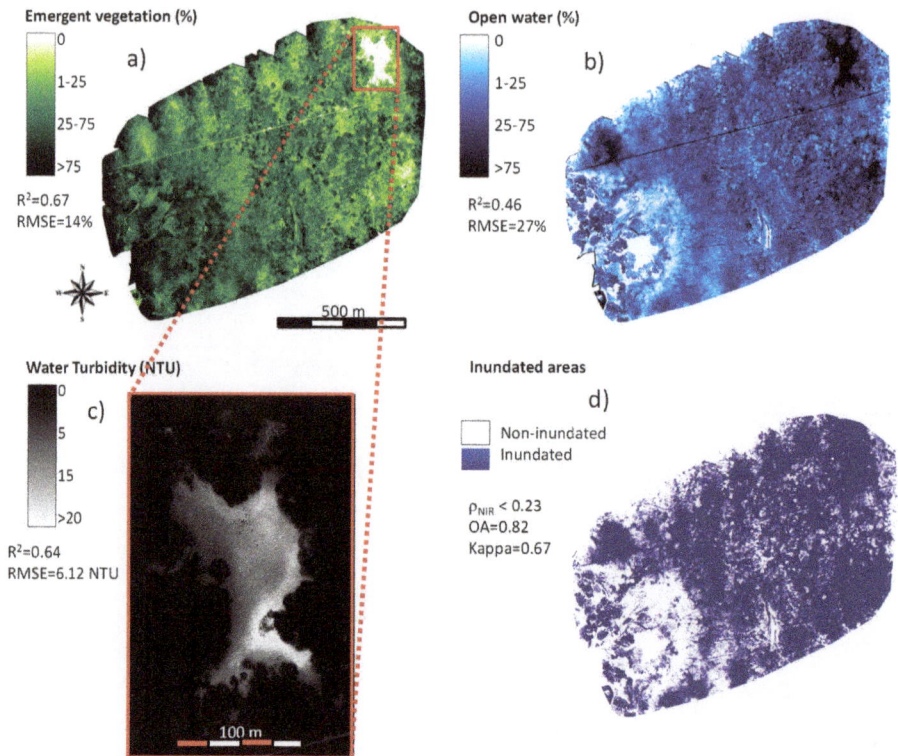

Figure 6. Resulting maps of (**a**) percent cover of emergent vegetation, (**b**) percentage of open water (**c**) water turbidity for the open water area found inside the red square and (**d**) inundation for the overflown area in Doñana marshes.

Regression tree optimally converged to 0.23 as the optimal NIR (Sequoia band 4) reflectance threshold to discriminate inundated from non-inundated areas (Figure 6).

While assessing spectral signatures of the most common aquatic plant species, most of them may easily be separated (Figure 7), specifically emergent aquatic plants such as grasses or saltmarsh bulrush (*Bolboschoenus maritimus*) versus floating aquatic species as *Ranunculus peltatus*. However, the less abundant species such as *Eleocharis palustris* and *Damasonium alisma* show spectral confusion with saltmarsh bulrush. The presence of the alien species *Azolla filiculoides* under saltmarsh bulrush canopy did not change the average reflectance of the helophyte alone.

According to separability analysis the best bands to discriminate among all the different pairs of species comparisons were the red edge band (Average $Z = 3.03$) and NIR band (Average $Z = 2.27$). The best discrimination was found between *Ranunculus* and grasses and between *Ranunculus* and the bulrush (average $Z = 3.99$ and average $Z = 3.16$). The less informative band was found to be the red band (average $Z = 1.06$) and the most difficult pair of species to be discriminated among each other were bulrush and bulrush with *Azolla*, and bulrush with *Azolla* and *Ranunculus*.

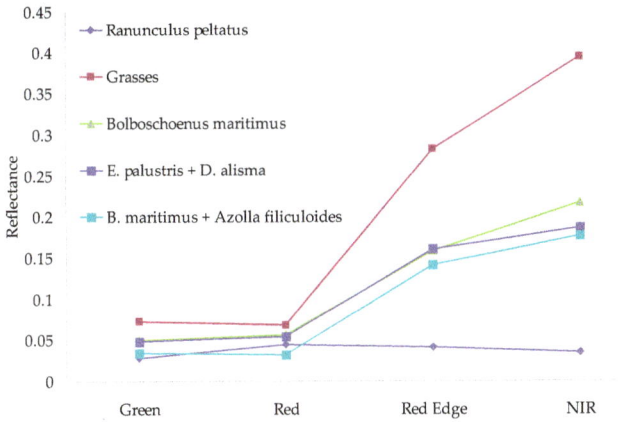

Figure 7. Mean spectral signatures of the different aquatic plant species found in the Doñana study area.

4.3. Reflectance Comparison with S2 Images

Visual assessment of overlay between S2 images and Sequoia spectral orthomosaics showed good color agreement both for Doñana and Braila study areas (Figure 4). Some linear elements, such as fences and buildings, were used to assess geometric matching which was always below one S2 pixel size.

Table 4 shows the values of the coefficient of determination (R^2) calculated band to band between Sequoia and the corresponding S2 spectral bands. RMSE are also provided for comparisons purposes.

Table 4. Overall band-by-band R^2 and RMSE values (% reflectance units) between the spectrally similar Sequoia and S2 bands.

	GREEN	RED	RED EDGE	NIR
R^2	0.68	0.61	0.65	0.70
RMSE	0.06	0.04	0.03	0.05

Although most of the bands show very high and positive correlation, there are still some discrepancies between both sensors. Consistently RMSE values are low, so that reflectances from both sensors can be compared despite having different spectral resolution.

4.4. Upscaling of Ground-Truth Information from Inundation to S2 Images

After the application of segmentation to the multispectral Sequoia orthomosaic we extended the ground-truth point information to the segments containing such points. This procedure enlarged the sampling area from 0.51 ha (51 points assigned by location to 51 S2 pixels used for modeling) up to 14.47 ha. We tested the classification of inundated areas by estimating the optimal threshold using the regression tree method, this time with the equivalent NIR band for S2, i.e., band 7. Reflectance values lower than 0.1964 were classified as inundated for the Doñana marsh area in the S2 scene (Figure 8). Overall agreement was 74% and kappa index value was 0.48.

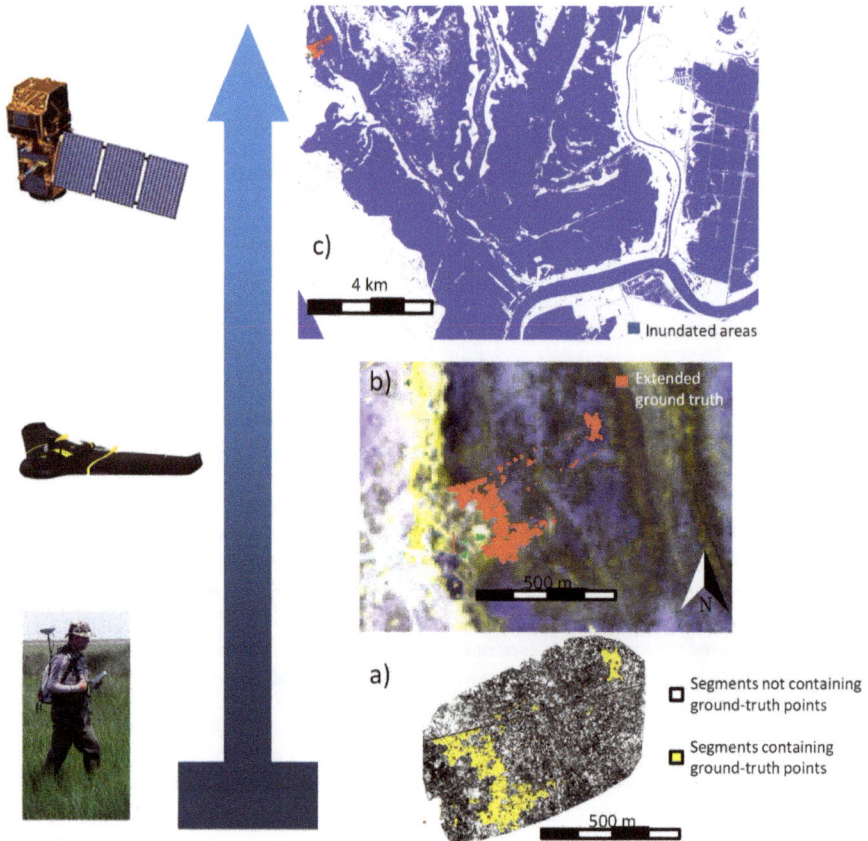

Figure 8. Upscaling process followed for the inundation mapping in Doñana LTSER wetlands. (a) Segments map resulting from multispectral Sequoia image showing those intercepted by ground-truth points (yellow polygons). (b) Location of selected segments (red extrapolated as inundated and green as non-inundated from ground-truth points) on top of RGB composite from S2 bands 7, 6, and 4. (c) Resulting inundation map for Doñana marshes in the S2 scene. eBee picture courtesy of Sensefly from Parrot Group®. Sentinel-2 image downloaded from Wikimedia Commons.

5. Discussion

Our work has demonstrated the applicability of multispectral mapping by UAVs in retrieving interesting remote sensing (RS) indicators relevant to assess ecological integrity (EI) in wetlands. Although a detailed set of EI indicators have been proposed to be used in LTER sites, here we suggest a set of practical remote sensing indicators and variables that can be used to evaluate wetlands condition by means of UAVs. The two studied wetlands are subject to similar drivers and pressures leading to changes in inundation level, hydroperiod or water quality [82]. Linked to these variables, percent plant cover of aquatic plants or floodplain trees is also a good EI indicator informing on both biotic diversity and within habitat diversity. In our case, the rapid assessment of inundation, water turbidity, percent plant cover and percentage of open water was very useful for managers in decision making in relation to *Azolla filiculoides* distribution which needs to be periodically assessed [29,83].

Aside from the good performance of UAV mapping, the study also demonstrates the rapid operation provided by this type of integrated solutions such as SenseFly eBee equipped with Parrot Sequoia multispectral camera. Mission planning can be designed just the same day of the flight and modified according to weather conditions, specifically wind direction and speed. The only constraint being the necessity of open and clear areas for landing which became evident while flying in the Small Island of Braila. Its low weight and manageability make of it very easy equipment to be transported even from one country to another. In addition, geometric correction can be satisfactory in the cases where no ground control points can be set, such as wetlands. The use of ground control points representative of large homogeneous areas reduces the effect of geometric error on thematic accuracy [67].

Through the many empirical modeling we found significant linear relationships between Sequoia spectral bands and indices and several RS wetlands indicators. The most informative band was band 4 in the NIR region. In general, Red and Green Sequoia bands were not informative for any of the analyses what might be linked to the saturation issues found for these bands by González-Piqueras et al. [84]. NIR was successfully related to percent open water and water depth while the spectral index NDRE was highly correlated with percentage of aquatic emergent plant cover, as reported in few other studies [85,86] but very weakly with water depth. Aquatic emergent plant height did only show low significant fits with the green band and not with the vegetation height model (VHM). Although VHM has been shown to work for crops [76], further investigation is required to enhance the application on aquatic vegetation. Aquatic emergent plants can be found with different plant densities and percent cover what might affect plant height retrieval from image matching [77]. Water turbidity was also significantly modeled through a linear relationship with the spectral index NDWI as already suggested by McFeeters [47] when it was proposed. However, only accurate results were found while applied on inundated areas with percentage open water higher than 50%. We used these models to generate the mapping for the remote sensing wetland indicators over the whole overflown areas (Figure 6).

Few aquatic plant species showed high spectral separability mainly using Red Edge and NIR bands [56]. The spectral signature of floating macrophytes was clearly separable from helophytes and grasses. However, the mapping of *Azolla filiculoides*, the alien invasive species in the understorey of bulrush was not noticeable for Sequoia bands. Previous studies have shown the difficulty of mapping aquatic ferns under a canopy of bulrush but the capability of identifying them while floating on open waters [15,29]. Although our research shows a high spectral separability between emergent and floating species, further investigation has to be done in examining the relationship of these practical RS wetlands indicators with the Essential Biodiversity Variables and Ecosystem Services [87].

Another relevant role played by UAV multispectral mapping evidenced in our work is the feasibility of enlarging the set of ground-truth data in such inaccessible or remote areas. By overflying areas where few ground points are sampled, based on spectral relationships as shown here, we can easily expand the ground-truth area to the whole flown area. Thus, we may easily increase sampling size for empirical modeling with satellite remote sensing imagery. In this study, we used the segmentation of multispectral orthomosaic from the UAV flight to enlarge ground-truth from the sampling points to the segments containing them. We only applied this upscaling for inundation mapping with Sentinel-2 images according to management request, although other RS wetland indicators might be upscaled too. The procedure allowed to accurately mapping inundation occurrence in Doñana marshes. While many upscaling studies have been carried out for crops [12,88] only few have been tested for natural vegetation [14,89,90] and none for aquatic vegetation. A different approach might be applied as well by directly using pixel values from the mapped wetland EI indicators as ground-truth data for the S2 scene. However, an estimation of the propagation error would be desirable for such an option as well as and assessment of the optimal resampling method from Sequoia to S2 pixels.

While scaling up ground-truth information, attention should be paid in the choice of the atmospheric correction model to be applied on satellite images. For instance, here we selected QGIS-SAC model based on DOS which may result in reflectance bias in comparison to other methods such as physical modeling or pseudo-invariant areas [91]. Here we found a close band-to-band relationship between Sequoia and S2 reflectances, but still a 40% unexplained variance that might be due to such biases or to the differences in spectral resolution [92]. These findings have necessarily to be refined by using handheld spectroradiometers in coincidence with the flight campaigns and satellite image acquisition.

We also want to emphasize the possibility to add temporal dimension to this kind of rapid assessments. Any mission might be repeated with the same planning during the inundation period or after sudden changes. These frequent missions will clearly enhance the data to model other relevant RS wetlands indicators such as hydroperiod [67]. In addition, recovery processes after disturbances can also be evidenced using a time series of UAV multispectral images [93,94].

6. Conclusions

In this paper, we confirmed the valuable and fast applicability of multispectral images captured by UAVs over two LTSER wetland sites in providing detailed mapping of inundation levels, water turbidity, and depth as well as aquatic plant cover. The resulting maps can play as detailed inputs to assess habitat diversity, wetlands dynamics, and ecosystem productivity and updated as frequently requested by managers or scientists. UAVs can easily reach remote areas with short-time flights resulting in an enlargement of the surveyed area (while allowed by local authorities). This advantage can definitely contribute to increase the area for ground-truth purposes when used for upscaling to satellite images. We tested the coherence between Sentinel-2 MSI bands and Parrot Sequoia showing the close multispectral relationship what makes possible the transference of UAV-scale to satellite-scale models.

Author Contributions: Conceptualization, R.D.-D. and M.A.; Formal analysis, R.D.-D.; Funding acquisition, R.D.-D. and M.A.; Investigation, R.D.-D. and M.A.; Methodology, R.D.-D.; Project administration, R.D.-D. and M.A.; Resources, C.C. and M.A.; Software, R.D.-D.; Supervision, M.A.; Validation, R.D.-D. and C.C.; Writing—original draft, R.D.-D.; Writing—review & editing, R.D.-D., C.C. and M.A.

Funding: The authors have received funding from the European Union's Horizon 2020 research and innovation program under grant agreements no. 654359 (eLTER Horizon 2020 project) and no. 641762 (ECOPOTENTIAL Horizon 2020 project). The authors have also received funding from the Romanian Space Agency under grant agreement 145/20.07.2017 "Integrated platform for monitoring of aquatic and terrestrial ecosystems based on in situ and satellite measurements" (RISE).

Acknowledgments: We are grateful to Espacio Natural de Doñana and Braila Island managers who provided permits for fieldwork in the protected areas with restricted access and to David Aragonés and Marco Antonio Espinosa who carried out the field sampling.

Conflicts of Interest: The authors declare no conflict of interest. The founding sponsors had no role in the design of the study; in the collection, analyses, or interpretation of data; in the writing of the manuscript, or in the decision to publish the results.

References

1. Díaz-Delgado, R.; Hurford, C.; Lucas, R. *Introducing the Book "The Roles of Remote Sensing in Nature Conservation." In The Roles of Remote Sensing in Nature Conservation*; Springer: Cham, Switzerland, 2017; pp. 3–10. ISBN 978-3-319-64330-4.
2. Ripple, W.J.; Wolf, C.; Newsome, T.M.; Galetti, M.; Alamgir, M.; Crist, E.; Mahmoud, M.I.; Laurance, W.F. World Scientists' Warning to Humanity: A Second Notice. *BioScience* **2017**, *67*, 1026–1028. [CrossRef]
3. Vaughan, H.; Brydges, T.; Fenech, A.; Lumb, A. Monitoring long-term ecological changes through the ecological monitoring and assessment network: Science-based and policy relevant. *Environ. Monit. Assess.* **2001**, *67*, 3–28. [CrossRef] [PubMed]
4. Díaz-Delgado, R. An Integrated Monitoring Programme for Doñana Natural Space: The Set-Up and Implementation. In *Conservation Monitoring in Freshwater Habitats*; Springer: Dordrecht, The Netherlands; Heidelberg, Germany; London, UK; New York, NY, USA, 2010; pp. 325–337. ISBN 978-1-4020-9278-7.
5. Haase, P.; Tonkin, J.D.; Stoll, S.; Burkhard, B.; Frenzel, M.; Geijzendorffer, I.R.; Häuser, C.; Klotz, S.; Kühn, I.; McDowell, W.H.; et al. The next generation of site-based long-term ecological monitoring: Linking essential biodiversity variables and ecosystem integrity. *Sci. Total Environ.* **2018**, *613–614*, 1376–1384. [CrossRef] [PubMed]

6. Müller, F.; Hoffmann-Kroll, R.; Wiggering, H. Indicating ecosystem integrity—Theoretical concepts and environmental requirements. *Ecol. Model.* **2000**, *130*, 13–23. [CrossRef]
7. Müller, F.; Gnauck, A.; Wenkel, K.-O.; Schubert, H.; Bredemeier, M. Theoretical Demands for Long-Term Ecological Research and the Management of Long-Term Data Sets. In *Long-Term Ecological Research*; Springer: Dordrecht, The Netherlands, 2010; pp. 11–25. ISBN 978-90-481-8781-2.
8. Singh, S.J.; Haberl, H.; Chertow, M.; Mirtl, M.; Schmid, M. *Long Term Socio-Ecological Research: Studies in Society-Nature Interactions Across Spatial and Temporal Scales*; Springer Science & Business Media: Dordrecht, The Netherlands, 2012; ISBN 978-94-007-1177-8.
9. Mirtl, M.; Orenstein, D.E.; Wildenberg, M.; Peterseil, J.; Frenzel, M. *Development of LTSER Platforms in LTER-Europe: Challenges and Experiences in Implementing Place-Based Long-Term Socio-ecological Research in Selected Regions*; Springer: Dordrecht, The Netherlands, 2013; ISBN 978-94-007-1176-1.
10. De Groot, R.S.; Alkemade, R.; Braat, L.; Hein, L.; Willemen, L. Challenges in integrating the concept of ecosystem services and values in landscape planning, management and decision making. *Ecol. Complex.* **2010**, *7*, 260–272. [CrossRef]
11. Lucas, R.; Díaz-Delgado, R.; Hurford, C. Expected Advances in a Rapidly Developing Work Area. In *The Roles of Remote Sensing in Nature Conservation*; Springer: Cham, Switzerland, 2017; pp. 309–318. ISBN 978-3-319-64330-4.
12. Zarco-Tejada, P.J.; Miller, J.R.; Noland, T.L.; Mohammed, G.H.; Sampson, P.H. Scaling-up and model inversion methods with narrowband optical indices for chlorophyll content estimation in closed forest canopies with hyperspectral data. *IEEE Trans. Geosci. Remote Sens.* **2001**, *39*, 1491–1507. [CrossRef]
13. Manfreda, S.; McCabe, M.; Miller, P.; Lucas, R.; Pajuelo Madrigal, V.; Mallinis, G.; Ben Dor, E.; Helman, D.; Estes, L.; Ciraolo, G.; et al. On the Use of Unmanned Aerial Systems for Environmental Monitoring. *Remote Sens.* **2018**, *10*, 641. [CrossRef]
14. Laliberte, A.S.; Goforth, M.A.; Steele, C.M.; Rango, A. Multispectral Remote Sensing from Unmanned Aircraft: Image Processing Workflows and Applications for Rangeland Environments. *Remote Sens.* **2011**, *3*, 2529–2551. [CrossRef]
15. Díaz-Delgado, R. Long-Term Ecological Monitoring at the Landscape Scale for Nature Conservation: The Example of Doñana Protected Area. In *The Roles of Remote Sensing in Nature Conservation*; Springer: Cham, Switzerland, 2017; pp. 65–76. ISBN 978-3-319-64330-4.
16. Lumbierres, M.; Méndez, P.F.; Bustamante, J.; Soriguer, R.; Santamaría, L. Modeling Biomass Production in Seasonal Wetlands Using MODIS NDVI Land Surface Phenology. *Remote Sens.* **2017**, *9*, 392. [CrossRef]
17. Green, A.J.; Alcorlo, P.; Peeters, E.T.; Morris, E.P.; Espinar, J.L.; Bravo-Utrera, M.A.; Bustamante, J.; Díaz-Delgado, R.; Koelmans, A.A.; Mateo, R.; et al. Creating a safe operating space for wetlands in a changing climate. *Front. Ecol. Environ.* **2017**, *15*, 99–107. [CrossRef]
18. Brouwer, R.; Bliem, M.; Getzner, M.; Kerekes, S.; Milton, S.; Palarie, T.; Szerényi, Z.; Vadineanu, A.; Wagtendonk, A. Valuation and transferability of the non-market benefits of river restoration in the Danube river basin using a choice experiment. *Ecol. Eng.* **2016**, *87*, 20–29. [CrossRef]
19. Schwarz, U. Hydromorphology of the Danube. In *The Danube River Basin*; The Handbook of Environmental Chemistry; Springer: Berlin/Heidelberg, Germany, 2014; pp. 469–479. ISBN 978-3-662-47738-0.
20. Dick, J.; Orenstein, D.E.; Holzer, J.; Wohner, C.; Achard, A.-L.; Andrews, C.; Avriel-Avni, N.; Beja, P.; Blond, N.; Cabello, J.; et al. What is socio-ecological research delivering? A literature survey across 25 international LTSER platforms. *Sci. Total Environ.* **2018**, *622–623*, 1225–1240. [CrossRef] [PubMed]
21. Pereira, H.M.; Ferrier, S.; Walters, M.; Geller, G.N.; Jongman, R.H.G.; Scholes, R.J.; Bruford, M.W.; Brummitt, N.; Butchart, S.H.M.; Cardoso, A.C.; et al. Essential Biodiversity Variables. *Science* **2013**, *339*, 277–278. [CrossRef] [PubMed]
22. Pettorelli, N.; Wegmann, M.; Skidmore, A.; Mücher, S.; Dawson, T.P.; Fernandez, M.; Lucas, R.; Schaepman, M.E.; Wang, T.; O'Connor, B.; et al. Framing the concept of satellite remote sensing essential biodiversity variables: Challenges and future directions. *Remote Sens. Ecol. Conserv.* **2016**, *2*, 122–131. [CrossRef]
23. Vihervaara, P.; Auvinen, A.-P.; Mononen, L.; Törmä, M.; Ahlroth, P.; Anttila, S.; Böttcher, K.; Forsius, M.; Heino, J.; Heliölä, J.; et al. How Essential Biodiversity Variables and remote sensing can help national biodiversity monitoring. *Glob. Ecol. Conserv.* **2017**, *10*, 43–59. [CrossRef]

24. Fritz, C.; Dörnhöfer, K.; Schneider, T.; Geist, J.; Oppelt, N. Mapping Submerged Aquatic Vegetation Using RapidEye Satellite Data: The Example of Lake Kummerow (Germany). *Water* **2017**, *9*, 510. [CrossRef]

25. Belluco, E.; Camuffo, M.; Ferrari, S.; Modenese, L.; Silvestri, S.; Marani, A.; Marani, M. Mapping salt-marsh vegetation by multispectral and hyperspectral remote sensing. *Remote Sens. Environ.* **2006**, *105*, 54–67. [CrossRef]

26. Hirano, A.; Madden, M.; Welch, R. Hyperspectral image data for mapping wetland vegetation. *Wetlands* **2003**, *23*, 436–448. [CrossRef]

27. Rodríguez-González, P.M.; Albuquerque, A.; Martínez-Almarza, M.; Díaz-Delgado, R. Long-term monitoring for conservation management: Lessons from a case study integrating remote sensing and field approaches in floodplain forests. *J. Environ. Manage.* **2017**. [CrossRef] [PubMed]

28. Townsend, P.A.; Walsh, S.J. Remote sensing of forested wetlands: Application of multitemporal and multispectral satellite imagery to determine plant community composition and structure in southeastern USA. *Plant Ecol.* **2001**, *157*, 129–149. [CrossRef]

29. Espinar, J.L.; Diaz-Delgado, R.; Bravo-Utrera, M.Á.; Vilà, M. Linking Azolla filiculoides invasion to increased winter temperatures in the Doñana marshland (SW Spain). *Aquat. Invasions* **2015**, *10*, 17–24. [CrossRef]

30. Díaz-Delgado, R.; Aragonés, D.; Ameztoy, I.; Bustamante, J. Monitoring marsh dynamics through remote sensing. In *Conservation Monitoring in Freshwater Habitats*; Hurford, C., Schneider, M., Cowx, I., Eds.; Springer: Dordrecht, The Netherlands; Heidelberg, Germany; London, UK; New York, NY, USA, 2010; pp. 375–386.

31. Bustamante, J.; Aragonés, D.; Afán, I.; Luque, C.J.; Pérez-Vázquez, A.; Castellanos, E.M.; Díaz-Delgado, R. Hyperspectral Sensors as a Management Tool to Prevent the Invasion of the Exotic Cordgrass Spartina densiflora in the Doñana Wetlands. *Remote Sens.* **2016**, *8*, 1001. [CrossRef]

32. Hestir, E.L.; Khanna, S.; Andrew, M.E.; Santos, M.J.; Viers, J.H.; Greenberg, J.A.; Rajapakse, S.S.; Ustin, S.L. Identification of invasive vegetation using hyperspectral remote sensing in the California Delta ecosystem. *Remote Sens. Environ.* **2008**, *112*, 4034–4047. [CrossRef]

33. Díaz-Delgado, R.; Mañez, M.; Martínez, A.; Canal, D.; Ferrer, M.; Aragonés, D. Using UAVs to Map Aquatic Bird Colonies. In *The Roles of Remote Sensing in Nature Conservation*; Springer: Cham, Switzerland, 2017; pp. 277–291. ISBN 978-3-319-64330-4.

34. Sardà-Palomera, F.; Bota, G.; Padilla, N.; Brotons, L.; Sardà, F. Unmanned aircraft systems to unravel spatial and temporal factors affecting dynamics of colony formation and nesting success in birds. *J. Avian Biol.* **2017**, *48*, 1273–1280. [CrossRef]

35. Anderson, K.; Gaston, K.J. Lightweight unmanned aerial vehicles will revolutionize spatial ecology. *Front. Ecol. Environ.* **2013**, *11*, 138–146. [CrossRef]

36. Hodgson, A.; Kelly, N.; Peel, D. Unmanned Aerial Vehicles (UAVs) for Surveying Marine Fauna: A Dugong Case Study. *PLOS ONE* **2013**, *8*, e79556. [CrossRef] [PubMed]

37. Hodgson, J.C.; Baylis, S.M.; Mott, R.; Herrod, A.; Clarke, R.H. Precision wildlife monitoring using unmanned aerial vehicles. *Sci. Rep.* **2016**, *6*, 22574. [CrossRef]

38. Kerr, J.T.; Ostrovsky, M. From space to species: Ecological applications for remote sensing. *Trends Ecol. Evol.* **2003**, *18*, 299–305. [CrossRef]

39. Seoane, J.; Bustamante, J.; Díaz-Delgado, R. Are existing vegetation maps adequate to predict bird distributions? *Ecol. Model.* **2004**, *175*, 137–149. [CrossRef]

40. Seoane, J.; Bustamante, J.; Díaz-Delgado, R. Competing roles for landscape, vegetation, topography and climate in predictive models of bird distribution. *Ecol. Model.* **2004**, *171*, 209–222. [CrossRef]

41. Silva, T.S.F.; Costa, M.P.F.; Melack, J.M.; Novo, E.M.L.M. Remote sensing of aquatic vegetation: Theory and applications. *Environ. Monit. Assess.* **2008**, *140*, 131–145. [CrossRef] [PubMed]

42. Lopez-Sanchez, J.M.; Vicente-Guijalba, F.; Erten, E.; Campos-Taberner, M.; Garcia-Haro, F.J. Retrieval of vegetation height in rice fields using polarimetric SAR interferometry with TanDEM-X data. *Remote Sens. Environ.* **2017**, *192*, 30–44. [CrossRef]

43. DeFries, R.; Eshleman, K.N. Land-use change and hydrologic processes: A major focus for the future. *Hydrol. Process.* **2004**, *18*, 2183–2186. [CrossRef]

44. Long, C.M.; Pavelsky, T.M. Remote sensing of suspended sediment concentration and hydrologic connectivity in a complex wetland environment. *Remote Sens. Environ.* **2013**, *129*, 197–209. [CrossRef]

45. Bustamante, J.; Pacios, F.; Díaz-Delgado, R.; Aragonés, D. Predictive models of turbidity and water depth in the Doñana marshes using Landsat TM and ETM+ images. *J. Environ. Manage.* **2009**, *90*, 2219–2225. [CrossRef] [PubMed]

46. Yamagata, Y.; Wiegand, C.; Akiyama, T.; Shibayama, M. Water turbidity and perpendicular vegetation indices for paddy rice flood damage analyses. *Remote Sens. Environ.* **1988**, *26*, 241–251. [CrossRef]

47. McFeeters, S.K. The use of the Normalized Difference Water Index (NDWI) in the delineation of open water features. *Int. J. Remote Sens.* **1996**, *17*, 1425–1432. [CrossRef]

48. Ritchie, J.C.; Zimba, P.V.; Everitt, J.H. Remote Sensing Techniques to Assess Water Quality. *Photogramm. Eng. Remote Sens.* **2003**, *69*, 695–704. [CrossRef]

49. Torgersen, C.E.; Faux, R.N.; McIntosh, B.A.; Poage, N.J.; Norton, D.J. Airborne thermal remote sensing for water temperature assessment in rivers and streams. *Remote Sens. Environ.* **2001**, *76*, 386–398. [CrossRef]

50. Jiménez-Muñoz, J.C.; Sobrino, J.A. A generalized single-channel method for retrieving land surface temperature from remote sensing data. *J. Geophys. Res. Atmospheres* **2003**, *108*, 4688. [CrossRef]

51. Bisht, G.; Venturini, V.; Islam, S.; Jiang, L. Estimation of the net radiation using MODIS (Moderate Resolution Imaging Spectroradiometer) data for clear sky days. *Remote Sens. Environ.* **2005**, *97*, 52–67. [CrossRef]

52. Ozesmi, S.L.; Bauer, M.E. Satellite remote sensing of wetlands. *Wetl. Ecol. Manag.* **2002**, *10*, 381–402. [CrossRef]

53. Lu, B.; He, Y.; Liu, H.H.T. Mapping vegetation biophysical and biochemical properties using unmanned aerial vehicles-acquired imagery. *Int. J. Remote Sens.* **2018**, *39*, 5265–5287. [CrossRef]

54. Peñuelas, J.; Gamon, J.A.; Griffin, K.L.; Field, C.B. Assessing community type, plant biomass, pigment composition, and photosynthetic efficiency of aquatic vegetation from spectral reflectance. *Remote Sens. Environ.* **1993**, *46*, 110–118. [CrossRef]

55. Ma, J.; Song, K.; Wen, Z.; Zhao, Y.; Shang, Y.; Fang, C.; Du, J. Spatial Distribution of Diffuse Attenuation of Photosynthetic Active Radiation and Its Main Regulating Factors in Inland Waters of Northeast China. *Remote Sens.* **2016**, *8*, 964. [CrossRef]

56. Adam, E.; Mutanga, O.; Rugege, D. Multispectral and hyperspectral remote sensing for identification and mapping of wetland vegetation: A review. *Wetl. Ecol. Manag.* **2010**, *18*, 281–296. [CrossRef]

57. Giardino, C.; Pepe, M.; Brivio, P.A.; Ghezzi, P.; Zilioli, E. Detecting chlorophyll, Secchi disk depth and surface temperature in a sub-alpine lake using Landsat imagery. *Sci. Total Environ.* **2001**, *268*, 19–29. [CrossRef]

58. Hestir, E.L.; Brando, V.E.; Bresciani, M.; Giardino, C.; Matta, E.; Villa, P.; Dekker, A.G. Measuring freshwater aquatic ecosystems: The need for a hyperspectral global mapping satellite mission. *Remote Sens. Environ.* **2015**, *167*, 181–195. [CrossRef]

59. Knox, S.H.; Dronova, I.; Sturtevant, C.; Oikawa, P.Y.; Matthes, J.H.; Verfaillie, J.; Baldocchi, D. Using digital camera and Landsat imagery with eddy covariance data to model gross primary production in restored wetlands. *Agric. For. Meteorol.* **2017**, *237–238*, 233–245. [CrossRef]

60. Kang, X.; Hao, Y.; Cui, X.; Chen, H.; Huang, S.; Du, Y.; Li, W.; Kardol, P.; Xiao, X.; Cui, L. Variability and Changes in Climate, Phenology, and Gross Primary Production of an Alpine Wetland Ecosystem. *Remote Sens.* **2016**, *8*, 391. [CrossRef]

61. He, T.; Liang, S.; Wang, D.; Chen, X.; Song, D.-X.; Jiang, B. Land Surface Albedo Estimation from Chinese HJ Satellite Data Based on the Direct Estimation Approach. *Remote Sens.* **2015**, *7*, 5495–5510. [CrossRef]

62. Zhang, K.; Kimball, J.S.; Running, S.W. A review of remote sensing based actual evapotranspiration estimation. *Wiley Interdiscip. Rev. Water* **2016**, *3*, 834–853. [CrossRef]

63. Kutser, T.; Paavel, B.; Verpoorter, C.; Ligi, M.; Soomets, T.; Toming, K.; Casal, G. Remote Sensing of Black Lakes and Using 810 nm Reflectance Peak for Retrieving Water Quality Parameters of Optically Complex Waters. *Remote Sens.* **2016**, *8*, 497. [CrossRef]

64. Oyama, Y.; Matsushita, B.; Fukushima, T. Distinguishing surface cyanobacterial blooms and aquatic macrophytes using Landsat/TM and ETM+ shortwave infrared bands. *Remote Sens. Environ.* **2015**, *157*, 35–47. [CrossRef]

65. Zhang, M.; Ustin, S.L.; Rejmankova, E.; Sanderson, E.W. Monitoring Pacific coast salt marshes using remote sensing. *Ecol. Appl.* **1997**, *7*, 1039–1053. [CrossRef]

66. DeVries, B.; Huang, C.; Lang, M.W.; Jones, J.W.; Huang, W.; Creed, I.F.; Carroll, M.L. Automated Quantification of Surface Water Inundation in Wetlands Using Optical Satellite Imagery. *Remote Sens.* **2017**, *9*, 807. [CrossRef]

67. Díaz-Delgado, R.; Aragonés, D.; Afán, I.; Bustamante, J. Long-Term Monitoring of the Flooding Regime and Hydroperiod of Doñana Marshes with Landsat Time Series (1974–2014). *Remote Sens.* **2016**, *8*, 775.
68. Jung, M.; Reichstein, M.; Ciais, P.; Seneviratne, S.I.; Sheffield, J.; Goulden, M.L.; Bonan, G.; Cescatti, A.; Chen, J.; de Jeu, R.; et al. Recent decline in the global land evapotranspiration trend due to limited moisture supply. *Nature* **2010**, *467*, 951. [CrossRef]
69. Pádua, L.; Vanko, J.; Hruška, J.; Adão, T.; Sousa, J.J.; Peres, E.; Morais, R. UAS, sensors, and data processing in agroforestry: A review towards practical applications. *Int. J. Remote Sens.* **2017**, *38*, 2349–2391. [CrossRef]
70. Franklin, S.E.; Ahmed, O.S.; Williams, G. Northern Conifer Forest Species Classification Using Multispectral Data Acquired from an Unmanned Aerial Vehicle. *Photogramm. Eng. Remote Sens.* **2017**, *83*, 501–507. [CrossRef]
71. Shen, Y.-Y.; Cattau, M.; Borenstein, S.; Weibel, D.; Frew, E.W. Toward an Architecture for Subalpine Forest Health Monitoring Using Commercial Off-the-Shelf Unmanned Aircraft Systems and Sensors. In Proceedings of the 17th AIAA Aviation Technology, Integration, and Operations Conference, Denver, CO, USA, 5 June 2017.
72. Ahmed, O.S.; Shemrock, A.; Chabot, D.; Dillon, C.; Williams, G.; Wasson, R.; Franklin, S.E. Hierarchical land cover and vegetation classification using multispectral data acquired from an unmanned aerial vehicle. *Int. J. Remote Sens.* **2017**, *38*, 2037–2052. [CrossRef]
73. Unger, J.; Reich, M.; Heipke, C. UAV-based photogrammetry: Monitoring of a building zone. *Int. Arch. Photogramm. Remote Sens.Spat. Inf. Sci.* **2014**, *XL*, 601–606. [CrossRef]
74. Tilling, A.K.; O'Leary, G.J.; Ferwerda, J.G.; Jones, S.D.; Fitzgerald, G.J.; Rodriguez, D.; Belford, R. Remote sensing of nitrogen and water stress in wheat. *Field Crops Res.* **2007**, *104*, 77–85. [CrossRef]
75. Rouse, J.W. *Monitoring the Vernal Advancement and Retrogradation (Green Wave Effect) of Natural Vegetation*; Technical report; Texas A&M Univ.; Remote Sensing Center: College Station, TX, USA, 1974.
76. Bendig, J.; Yu, K.; Aasen, H.; Bolten, A.; Bennertz, S.; Broscheit, J.; Gnyp, M.L.; Bareth, G. Combining UAV-based plant height from crop surface models, visible, and near infrared vegetation indices for biomass monitoring in barley. *Int. J. Appl. Earth Obs. Geoinformation* **2015**, *39*, 79–87. [CrossRef]
77. Ruzgienė, B.; Berteška, T.; Gečyte, S.; Jakubauskienė, E.; Aksamitauskas, V.Č. The surface modeling based on UAV Photogrammetry and qualitative estimation. *Measurement* **2015**, *73*, 619–627. [CrossRef]
78. Davis, S.M.; Landgrebe, D.A.; Phillips, T.L.; Swain, P.H.; Hoffer, R.M.; Lindenlaub, J.C.; Silva, L.F. *Remote Sensing: The Quantitative Approach*; McGraw-Hill International Book Co: New York, NY, USA, 1978; p. 405.
79. Congedo, L. Semi-Automatic Classification Plugin Documentation. Release 5.3.6.1. 2016. Available online: https://media.readthedocs.org/pdf/semiautomaticclassificationmanual-v5/latest/semiautomaticclassificationmanual-v5.pdf (accessed on 20 December 2018).
80. Chavez, P.S. Image-based atmospheric corrections-revisited and improved. *Photogramm. Eng. Remote Sens.* **1996**, *62*, 1025–1035.
81. Moran, M.S.; Jackson, R.D.; Slater, P.N.; Teillet, P.M. Evaluation of simplified procedures for retrieval of land surface reflectance factors from satellite sensor output. *Remote Sens. Environ.* **1992**, *41*, 169–184. [CrossRef]
82. Haberl, H.; Gaube, V.; Díaz-Delgado, R.; Krauze, K.; Neuner, A.; Peterseil, J.; Plutzar, C.; Singh, S.J.; Vadineanu, A. Towards an integrated model of socioeconomic biodiversity drivers, pressures and impacts. A feasibility study based on three European long-term socio-ecological research platforms. *Ecol. Econ.* **2009**, *68*, 1797–1812. [CrossRef]
83. Díaz-Delgado, R. Caso 5. La teledetección como herramienta en la cartografía de especies invasoras: Azolla filiculoides en Doñana. In *Invasiones Biológicas*; Vila, M., Valladares, F., Traveset, A., Santamaría, L., Castro, P., Eds.; Consejo Superior de Investigaciones Científicas: Madrid, Spain, 2008; pp. 159–163.
84. González-Piqueras, J.; Sánchez, S.; Villodre, J.; López, H.; Calera, A.; Hernández-López, D.; Sánchez, J.M. Radiometric Performance of Multispectral Camera Applied to Operational Precision Agriculture. In Proceedings of the IGARSS 2018—2018 IEEE International Geoscience and Remote Sensing Symposium, Valencia, Spain, 22–27 July 2018; 2018; pp. 3393–3396.
85. Cho, H.J.; Kirui, P.; Natarajan, H. Test of Multi-spectral Vegetation Index for Floating and Canopy-forming Submerged Vegetation. *Int. J. Environ. Res. Public. Health* **2008**, *5*, 477–483. [CrossRef]
86. Jakubauskas, M.; Kindscher, K.; Fraser, A.; Debinski, D.; Price, K.P. Close-range remote sensing of aquatic macrophyte vegetation cover. *Int. J. Remote Sens.* **2000**, *21*, 3533–3538. [CrossRef]

87. Alcaraz-Segura, D.; Bella, C.M.D.; Straschnoy, J.V. *Earth Observation of Ecosystem Services*, 1st ed.; CRC Press: Boca Raton, FL, USA, 2013; ISBN 978-1-4665-0588-9.

88. Jay, S.; Baret, F.; Dutartre, D.; Malatesta, G.; Héno, S.; Comar, A.; Weiss, M.; Maupas, F. Exploiting the centimeter resolution of UAV multispectral imagery to improve remote-sensing estimates of canopy structure and biochemistry in sugar beet crops. *Remote Sens. Environ.* **2018**. [CrossRef]

89. Fraser, R.H.; Olthof, I.; Lantz, T.C.; Schmitt, C. UAV photogrammetry for mapping vegetation in the low-Arctic. *Arct. Sci.* **2016**, *2*, 79–102. [CrossRef]

90. Nelson, P.; Paradis, D.P. Evaluating rapid ground sampling and scaling estimated plant cover using UAV imagery up to Landsat for mapping arctic vegetation. *AGU Fall Meet. Abstr.* **2017**, *21*. abstract #B21F-2016.

91. Padró, J.-C.; Pons, X.; Aragonés, D.; Díaz-Delgado, R.; García, D.; Bustamante, J.; Pesquer, L.; Domingo-Marimon, C.; González-Guerrero, Ó.; Cristóbal, J.; et al. Radiometric correction of simultaneously acquired Landsat-7/Landsat-8 and Sentinel-2A imagery using Pseudoinvariant Areas (PIA): Contributing to the Landsat time series legacy. *Remote Sens.* **2017**, *9*, 1319. [CrossRef]

92. Padró, J.-C.; Muñoz, F.-J.; Ávila, L.Á.; Pesquer, L.; Pons, X. Radiometric Correction of Landsat-8 and Sentinel-2A Scenes Using Drone Imagery in Synergy with Field Spectroradiometry. *Remote Sens.* **2018**, *10*, 1687. [CrossRef]

93. Turner, D.; Lucieer, A.; de Jong, S.M. Time Series Analysis of Landslide Dynamics Using an Unmanned Aerial Vehicle (UAV). *Remote Sens.* **2015**, *7*, 1736–1757. [CrossRef]

94. Suárez, L.; Zarco-Tejada, P.J.; González-Dugo, V.; Berni, J.A.J.; Sagardoy, R.; Morales, F.; Fereres, E. Detecting water stress effects on fruit quality in orchards with time-series PRI airborne imagery. *Remote Sens. Environ.* **2010**, *114*, 286–298. [CrossRef]

![drones logo] *drones*

MDPI

Article

Estimating Wildlife Tag Location Errors from a VHF Receiver Mounted on a Drone

André Desrochers [1,*], Junior A. Tremblay [2], Yves Aubry [3], Dominique Chabot [4], Paul Pace [5] and David M. Bird [6]

[1]   Département des Sciences du Bois et de la Forêt, Université Laval, 2405 rue de la Terrasse, Québec, QC G1V 0A6, Canada
[2]   Environment and Climate Change Canada, Science and Technology, 801-1550 avenue d'Estimauville, Québec, QC G1J 0C3, Canada; junior.tremblay@canada.ca
[3]   Environment and Climate Change Canada, Canadian Wildlife Service, 801-1550 avenue d'Estimauville, Québec, QC G1J 0C3, Canada; yves.aubry@canada.ca
[4]   droneMetrics, 7 Tauvette Street, Ottawa, ON K1B 3A1, Canada; dominique.chabot@mail.mcgill.ca
[5]   Defense Research and Development Canada, 3701 Carling Avenue, Ottawa, ON K1A 0Z4, Canada; paul.pace@rogers.com
[6]   Avian Science and Conservation Centre of McGill University, c/o 10980 Dunne Road, North Saanich, BC V8L 5J1, Canada; david.bird@mcgill.ca
*   Correspondence: andre.desrochers@sbf.ulaval.ca; Tel.: +1-418-656-2131

Received: 9 November 2018; Accepted: 7 December 2018; Published: 11 December 2018

Abstract: Recent studies have demonstrated the high potential of drones as tools to facilitate wildlife radio-tracking in rugged, difficult-to-access terrain. Without estimates of accuracy, however, data obtained from receivers attached to drones will be of limited use. We estimated transmitter location errors from a drone-borne VHF (very high frequency) receiver in a hilly and dense boreal forest in southern Québec, Canada. Transmitters and the drone-borne receiver were part of the Motus radio-tracking system, a collaborative network designed to study animal movements at local to continental scales. We placed five transmitters at fixed locations, 1–2 m above ground, and flew a quadrotor drone over them along linear segments, at distances to transmitters ranging from 20 m to 534 m. Signal strength was highest with transmitters with antennae pointing upwards, and lowest with transmitters with horizontal antennae. Based on drone positions with maximum signal strength, mean location error was 134 m (range 44–278 m, n = 17). Estimating peak signal strength against drone GPS coordinates with quadratic, least-squares regressions led to lower location error (mean = 94 m, range 15–275 m, n = 10) but with frequent loss of data due to statistical estimation problems. We conclude that accuracy in this system was insufficient for high-precision purposes such as finding nests. However, in the absence of a dense array of fixed receivers, the use of drone-borne Motus receivers may be a cost-effective way to augment the quantity and quality of data, relative to deploying personnel in difficult-to-access terrain.

Keywords: radio-tracking; Motus; drone; boreal forest; precision; accuracy; response surface; forêt Montmorency

1. Introduction

For several decades, radio-tracking has proven itself as a valuable tool for the investigation of animal movements at a wide range of temporal and spatial scales [1]. The advent of the Global Positioning System (GPS) and similar satellite-based systems has revolutionized wildlife tracking by transmitting precise coordinates directly from tagged animals. However, GPS transmitters remain costly and too heavy for small animals. As a result, conventional, non-GPS, transmitters remain the predominantly used tracking technology in studies of small songbirds and many other terrestrial applications. With the exception of light-sensitive geolocators, which require recaptures to retrieve data stored in memory [2], the location of non-GPS transmitters usually has to be inferred from signal detection, signal strength, compass direction or combinations of those.

To get precise estimates of the location of animals fitted with non-GPS transmitters, field biologists will often resort to triangulation or homing, i.e., getting progressively closer to the focal animal [3]. In rugged, difficult-to-access terrain, the latter two approaches are labor-intensive and may be well outside research budgets, not to mention ever-increasing concerns about safety [4]. For species ranging over kilometers or more, the use of conventional aircraft such as Cessna planes is often the only way to obtain sufficient amounts of data, but such campaigns also pose a safety risk, with aviation accidents determined to be the leading cause of mortality among wildlife workers in the United States from 1937–2000 [5].

The advent of drones in the civil sector may offer immense potential for combining affordability, safety, and accuracy in the effort to document movements of animals fitted with non-GPS transmitters. Chabot and Bird [6] provide a review of recent advances in the use of drones for wildlife applications. They point out that despite its potential, drone-borne wildlife radio-tracking remains underdeveloped, possibly due to enduring skepticism about its potential and/or technological and operational barriers. To this day, the published literature suggests that, with few exceptions, the subject continues to be largely approached as an engineering curiosity more so than an endeavor by those who stand to benefit from its development: wildlife researchers and managers [7–10].

One of the concerns that needs to be addressed to promote the effective use of drones in wildlife radio-tracking is accuracy. Signal power density is proportional to the inverse square of the distance between the transmitter and the receiver. In principle thus, knowing the strength and direction of a transmitter signal from two locations, sufficiently distinct in space and sufficiently close in time, should yield highly accurate positions. However, transmitter signals are dampened by trees, rocks, etc., to varying extents, and may exhibit multi-path effects, making it practically impossible to infer transmitter locations from two locations with intervening obstacles.

We estimated transmitter location errors with fixed-location "test" transmitters and a drone-borne receiver in a hilly and dense forest composed mostly of balsam fir (*Abies balsamea*). We used a simple quadratic, two-dimensional response surface of signal strength against Latitude and Longitude.

2. Materials and Methods

We conducted this study in September 2016 at Forêt Montmorency (47.4 N, 71.1 W), a teaching and research forest located north of Québec City, Québec, Canada. The study area is a dense balsam fir/white birch (*Betula papyrifera*) boreal forest landscape with altitudes ranging from 750 to 1000 m, covered by a dense network of forestry roads (for details see [11]). We conducted seven flights, within two sectors each covering ~0.2 km^2, each composed of a matrix of old, mid-successional, and early-successional balsam fir stands resulting from clearcutting (Figure 1). Tree height in the study site varied from ~4 m to 12 m, with ~2500–10,000 stems/ha.

Each flight was performed by a custom-built heavy-lift quadrotor drone based on a Gryphon Dynamics airframe (Daegu, South Korea) and a Pixhawk flight controller (3D Robotics, Berkeley, CA, USA), with a payload capacity of about 4 kg including the battery. The drone was programmed to fly at a fixed altitude of 50 m above the ground at the location where it was launched and a forward speed of 5 m/s. We automated flights from takeoff to landing, and monitored them from the ground using a tablet computer. We mounted a radio receiver system on the ground-facing side of the drone. The radio receiver system was composed of a hanging omnidirectional dipole antenna attached to a weight at the bottom end, and coupled to a Funcube Pro+ dongle. The dongle was connected to a BeagleBone computer programmed to monitor and record signals simultaneously from multiple transmitters (for details see [12]).

Before each flight, we deployed five radio-transmitters at distances ranging from 18 m to 507 m from one another (Figure 1). We placed each transmitter in a tree at ca. 1.3 m above ground, with the antenna pointing up, down or horizontal. Transmitters were avian nanotags model NTQB-4-2, Lotek Wireless Inc., Newmarket, ON, Canada). Each nanotag had a unique set of pulses delivered each 5 s at a frequency of 166.38 MHz (VHF; very high frequency), a standard used by the Motus Wildlife Tracking System [12,13].

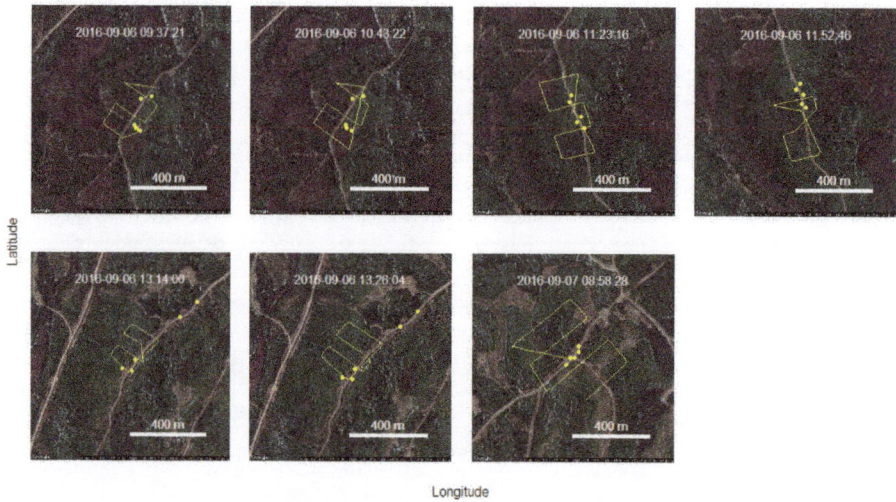

Figure 1. Flight transects and test transmitter locations, in yellow. Takeoff times are indicated.

After the completion of each flight, we downloaded data from two sources: the telemetry receiver and the drone navigation log. Timestamps from the two sources were synchronized, allowing us to match each signal reception from test transmitters to the exact location of the drone and in turn, the distance between the transmitters and the receiver.

As pointed out earlier, several factors may influence signal reception and strength. However, on average, signal strength should provide an unbiased estimate of the distance between the transmitter and the receiver. Thus, a two-dimensional array, i.e., a map of signal strengths, should inform us about the true location of the transmitter. We estimated the location errors with two methods. First, we retained the drone location at which the signal was strongest, and calculated the Euclidean distance between the drone XY coordinates, and the XY coordinates of the source. Second, we modelled signal strength as a function of the drone's XY coordinates in meters from a Modified Transverse Mercator map projection (Easting and Northing) using two quadratic functions:

$$\text{Easting: Dbm} \sim \beta_0 + \beta_1(\text{Easting}) + \beta_2(\text{Easting}^2) + \varepsilon \tag{1}$$

$$\text{Northing: Dbm} \sim \beta_0 + \beta_1(\text{Northing}) + \beta_2(\text{Northing}^2) + \varepsilon \tag{2}$$

where Dbm is the signal's strength, β_i regression estimates, and ε a vector of model residuals. The formulas yielded a peak signal strength when the regression estimate for the quadratic term was negative. Note that in the presence of a peak signal strength both on X and Y coordinates, only *relative* signal strength will be required to estimate transmitter locations. Differences in signal strength among transmitters, whether because of manufacturing or placement in the forest, are measured by the models' intercept (β_0). We obtained Easting and Northing estimates by:

$$\hat{E}, \hat{N} = \frac{-\beta_1}{2 \cdot \beta_2} \tag{3}$$

We conducted all analyses with the statistical software R version 3.5.0 [14].

3. Results

We obtained 669 detections of the test transmitters from the combined drone flights. Signal strength decreased significantly with increasing distance to drone, with the furthest detection at 534 m (Figure 2, $F_{1,646} = 181.7, p < 0.001$). The orientation of the transmitter's antenna also had a significant effect on signal strength ($F_{2,646} = 62.6, p < 0.001$), with antennae pointing upward performing best.

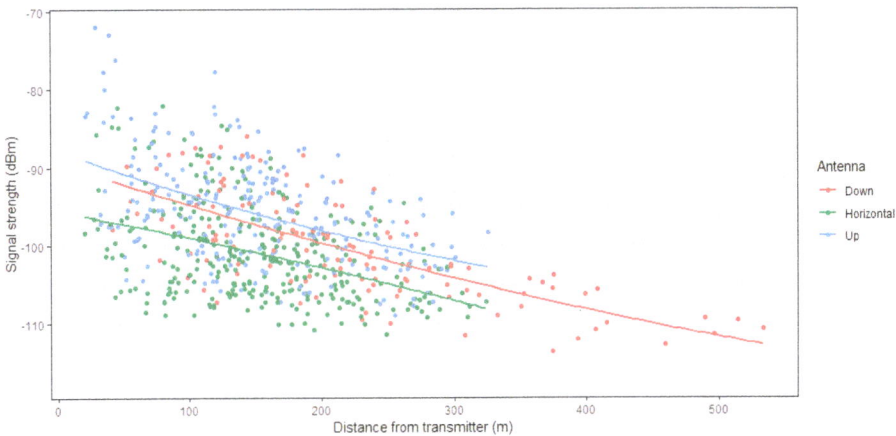

Figure 2. Signal strength in response to the distance between the drone and the transmitter.

Given that we conducted seven flights, each with five test transmitters deployed, we obtained 35 detection sets. Five to 42 detections were obtained for each detection set (mean = 19.9). Based solely on drone locations when signal strength was greatest, mean location error was 132.3 m (range: 28.8–294.2 m, n = 35). The reliability of the maximum strength method is questionable when the strongest signal is on the periphery of the drone route, because in those cases the transmitter was likely outside of the range covered by the drone. To prevent this, we removed all detection sets where the strongest signal came from a drone location on the periphery of the convex hull enclosing the detection set. The resulting subset of data yielded a mean location error of 134.0 m (range: 43.9–278.0 m, n = 17).

Nearest distances to transmitters yielded strongest signals in only two of the 35 detection sets. Furthermore, signal strength did not always increase nearer transmitters (Figure 3), leading to only seven cases where quadratic regression coefficients of signal strength against X or Y coordinates were negative, i.e., leading to a maximum estimated signal strength as required for position estimation.

Of the seven cases with estimable positions, we dropped one case with an estimate error (3034 m) greater than the maximum known distance between the drone and the transmitter (534 m). Figure 4 illustrates the remaining six cases where quadratic curve-fitting yielded estimable positions.

The mean location error from the quadratic method was 69.9 m (range 20.8–161.3 m).

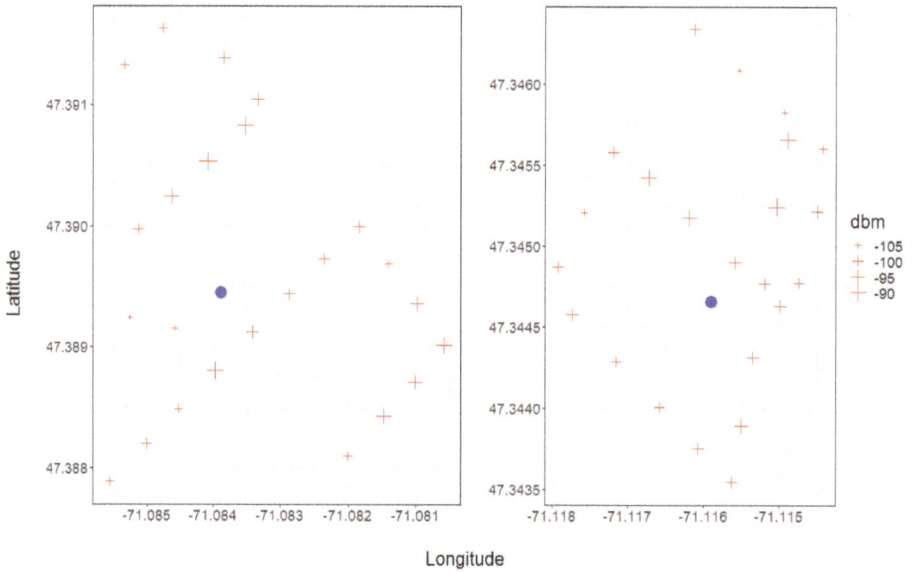

Figure 3. Left: best case scenario, with signal strength, depicted by cross size, roughly increasing toward the location of the test transmitter (blue dot). Right, worst case scenario, with signal strength showing no obvious relationship with distance to transmitter.

Figure 4. Transmitter location errors from quadratic regression estimation, based on signal strength. Test transmitter locations in yellow, estimated locations in red.

4. Discussion

The main outcome of this study was that drones offer an alternative to the more labor-intensive, traditional approaches for radio-tracking small birds, amphibians, or small mammals in rugged terrain. All five transmitters were detected on each of the seven flights, thus for the purposes of simple detection at a range of a few hundred meters, a drone appears highly effective. However, the precision of the detection-by-drone method is likely insufficient for finer-scale applications such as finding nests or dens or documenting microhabitat use. We explored two ways to estimate source locations, based on observed vs. modeled maximum signal strength. Observed maximum signal strength has the benefit of being simple and easily obtainable (larger sample), but it wastes the bulk of the information obtained by the drone. Furthermore, it is overly sensitive to outliers. The estimation method based on quadratic estimation has the advantage of combining a comprehensive use of the data with computational simplicity. In the present study, the quadratic method yielded disappointing results. However, we believe that this method should be assessed more thoroughly with denser flight paths, and a less variable elevation of the drone above ground, than in the present study. Additionally, field trials should be conducted over open areas such as fields to evaluate the statistical noise, and possible bias, caused by dense canopy. Our results were possibly influenced by the proximity of test transmitters to road edges, so future field trials could be more reliable if transmitters were placed at more varying distances from roads.

Given the aims of this study, it was natural to design the drone search pattern so that it would fly directly over the transmitters. Of course, in real searches for animals wearing transmitters, search patterns would result from a tradeoff between high density, e.g., the mean distance between flight segments, and extent. Maximum detection distance should set a lower limit to search pattern density. We found that signal strength decreased approximately linearly with increasing distance between the source and the drone. Given that we were able to detect few signals at distances further than 500 m, we believe that drones should always fly within 500 m of locations where animals wearing tags are expected, or test transmitters were placed, in the case of calibration studies such as the present one.

Even in cases where drones would be flown in a dense search pattern over large extents, locating tags carried by animals in motion will be more challenging than locating stationary tags. The proximity of an animal body is known to amplify signals [15], but fortunately this effect should not influence location estimation, which is based on *relative* signal strength from one drone location to another. Animals moving fast on the horizontal plane, e.g., birds in flight, would undoubtedly pose the greatest challenge. However, even animals remaining in fixed locations may prove more difficult to locate than fixed tags, if they move (e.g., foraging), because signal strength depends on tag angle relative to the drone, as we found here.

Over the past several years, there has been a dramatic increase in the use of drones to survey and monitor birds by means of optical imaging, including breeding colonies [16–18], wintering and migrating waterbirds [19–21], and individual nest inspections [22–24]. Rapid uptake of these applications has been made possible by the relative maturity, simplicity and accessibility of the requisite technologies, namely sophisticated and user-friendly drone flight control systems combined with compact and very high-resolution digital cameras. However, the applicability of these approaches remains limited to relatively large and/or unconcealed birds, whereas a considerable proportion of species under study or management are small and challenging to locate or directly observe [25]. Thus, it is of continuing interest to develop drone-based solutions to monitor and track birds using alternative sensing methods that do not rely on direct visual observation of subjects, including acoustic sensors [26] and radio telemetry. Such developments would help in addressing more elaborate questions, such as habitat selection. This also extends to non-avian species: Chabot and Bird [6] identified a range of wildlife taxa whose study and monitoring could potentially benefit from drone-based radio-tracking, such as small- to medium-sized mammals including primates, mustelids, rodents and bats, as well as lizards, snakes, land turtles and amphibians.

We believe that the lack of breakthrough in drone-based radio-tracking in the last decade results in part from a focus on more appealing, but more challenging, solutions. One of those is based on an "active" localization system whereby multiple antennas mounted on a drone would enable real-time, onboard triangulation of an animal's location, which in turn automatically adjusts the drone's heading to fly towards the animal and pinpoint its position [6]. Indeed, the complexities involved in bringing this idea to fruition have proven exacting, and the best working prototypes to date still require a human operator at the drone's ground control station to manually determine and transmit flight path adjustments based on real-time feedback from the onboard antenna-triangulation system [7]. In contrast, the approach detailed in our study and explored by few others [27] consists of a "passive" localization system whereby the drone executes a preprogrammed flight path over an area potentially containing one or more radio-tagged animals. The varying signal strength of a given transmitter received by a single antenna on the drone at multiple locations along its flight path is analyzed post hoc to estimate the animal's location. Although not as compelling a solution as active localization, this passive approach can be much more readily implemented using existing technology (i.e., a drone capable of autonomous waypoint navigation and a simple antenna–receiver–logger system) without needing to develop and integrate sophisticated new gadgetry.

Continuing technological and operational advancements of drones are likely to improve their effectiveness for wildlife radio-tracking going forward. Already, "terrain following" capabilities have now been integrated into most popular drone flight control systems, enabling the aircraft to maintain a constant altitude above ground level in areas of varying relief, and consequently better normalization of the strength of radio signals received from the ground. The flight range of drones is also currently limited both by battery capacity and, often more significantly, airspace regulations that predominantly restrict drone operation to within visual line of sight of operators on the ground [28,29]. This restriction tends to be especially crippling in forests, where tall and dense trees, and hilly topography surrounding ground operators cause them to quickly lose sight of the drone as it flies away. It is therefore promising that regulatory agencies in several countries have recently undertaken more serious considerations of allowing drone operations beyond visual line of sight (BVLOS) under certain conditions, and even begun granting BVLOS approvals in limited cases. Regarding flight endurance, fixed-wing drones can typically remain airborne significantly longer (upwards of an hour) than rotary-wing drones (typically <30 min), but the takeoff and landing space requirements of the former tend to prohibit their use in areas such as forests, whereas the latter feature more versatile vertical takeoff and landing (VTOL). However, a growing number of hybrid VTOL fixed-wing drones have recently begun to enter the commercial market.

5. Conclusions

Despite the limited number and extent of drone flights in this study, we were able to obtain detection sets with enough detail to provide an operational assessment of transmitter location errors. We conclude that accuracy in this system was insufficient for high-precision purposes such as finding nests. However, in the absence of a dense array of telemetry towers, the use of drone-borne receivers may be a cost-effective way to enhance the quantity and quality of data, relative to deploying personnel in difficult-to-access terrain.

Author Contributions: Conceptualization, A.D., J.A.T., Y.A., D.M.B. Methodology, A.D., J.A.T., Y.A., P.P., D.M.B., D.C.; Validation, A.D., J.A.T.; Formal Analysis, A.D.; Resources, A.D., J.A.T., Y.A., D.M.B., P.P.; Data Curation, A.D.; Writing-Original Draft Preparation, A.D.; Writing-Review & Editing, A.D., D.C., J.A.T., Y.A., D.M.B., P.P.; Project Administration, A.D., J.A.T.; Funding Acquisition, A.D., J.A.T., Y.A.

Funding: This research was funded by the Natural Sciences and Engineering Council of Canada grant number 170173 to Desrochers and Environment and Climate Change Canada.

Acknowledgments: We thank the personnel at the Forêt Montmorency for their significant support.

Conflicts of Interest: The authors declare no conflict of interest. The funders had no role in the design of the study; in the collection, analyses, or interpretation of data; in the writing of the manuscript, and in the decision to publish the results.

References

1. Millspaugh, J.J.; Marzluff, J.M. *Radio Tracking and Animal Populations*; Millspaugh, J., Marzluff, J., Eds.; Academic Press: New York, NY, USA, 2001; p. 474.
2. Stutchbury, B.J.; Tarof, S.A.; Done, T.; Gow, E.; Kramer, P.M.; Tautin, J.; Fox, J.W.; Afanasyev, V. Tracking long-distance songbird migration by using geolocators. *Science* **2009**, *323*, 896. [CrossRef] [PubMed]
3. White, G.C.; Garrott, R.A. *Analysis of Wildlife Radio-Tracking Data*; Academic Press: New York, NY, USA, 1990; p. xiii+383.
4. Thomas, B.; Holland, J.D.; Minot, E.O. Wildlife tracking technology options and cost considerations. *Wildl. Res.* **2011**, *38*, 653–663. [CrossRef]
5. Sasse, D.B. Job-related mortality of wildlife workers in the United States, 1937–2000. *Wildl. Soc. Bull.* **2003**, *31*, 1015–1020.
6. Chabot, D.; Bird, D.M. Wildlife research and management methods in the 21st century: Where do unmanned aircraft fit in? *J. Unmanned Veh. Syst.* **2015**, *3*, 137–155. [CrossRef]
7. Cliff, O.M.; Saunders, D.L.; Fitch, R. Robotic ecology: Tracking small dynamic animals with an autonomous aerial vehicle. *Sci. Robot.* **2018**, *3*, eaat8409. [CrossRef]
8. Webber, D.; Hui, N.; Kastner, R.; Schurgers, C. Radio receiver design for unmanned aerial wildlife tracking. In Proceedings of the 2017 International Conference on Computing, Networking and Communications, Santa Clara, CA, USA, 26–29 January 2017. [CrossRef]
9. VonEhr, K.; Hilaski, S.; Dunne, B.E.; Ward, J. Software defined radio for direction-finding in UAV wildlife tracking. In Proceedings of the 2016 IEEE International Conference on Electro Information Technology, Grand Forks, ND, USA, 19–21 May 2016; pp. 464–469. [CrossRef]
10. Bayram, H.; Doddapaneni, K.; Stefas, N.; Isler, V. Active localization of VHF collared animals with aerial robots. In Proceedings of the 2016 IEEE International Conference on Automation Science and Engineering, Fort Worth, TX, USA, 21–24 August 2016; pp. 934–939. [CrossRef]
11. Corbani, A.C.; Hachey, M.-H.; Desrochers, A. Food provisioning and parental status in songbirds: Can occupancy models be used to estimate nesting performance? *PLoS ONE* **2014**, *9*, e101765. [CrossRef]
12. Taylor, P.D.; Crewe, T.L.; Mackenzie, S.A.; Lepage, D.; Aubry, Y.; Crysler, Z.; Finney, G.; Francis, C.M.; Guglielmo, C.G.; Hamilton, D.J.; et al. The Motus Wildlife Tracking System: A collaborative research network to enhance the understanding of wildlife movement. *ACE-ECO* **2017**, *12*. [CrossRef]
13. Tremblay, J.; Desrochers, A.; Aubry, Y.; Pace, P.; Bird, D.M. A low-cost technique for radio-tracking wildlife using a small standard unmanned aerial vehicle. *J. Unmanned Veh. Syst.* **2017**, *5*, 102–108. [CrossRef]
14. R Core Team. *R: A Language and Environment for Statistical Computing*; R Foundation for Statistical Computing: Vienna, Austria, 2015. Available online: https://www.R-project.org/ (accessed on 22 November 2018).
15. Naef-Daenzer, B.; Früh, D.; Stalder, M.; Wetli, P.; Weise, E. Miniaturization (0.2 g) and evaluation of attachment techniques of telemetry transmitters. *J. Exp. Biol.* **2005**, *208*, 4063–4068. [CrossRef]
16. Borowicz, A.; McDowall, P.; Youngflesh, C.; Sayre-McCord, T.; Clucas, G.; Herman, R.; Forrest, S.; Rider, M.; Schwaller, M.; et al. Multi-modal survey of Adélie penguin mega-colonies reveals the Danger Islands as a seabird hotspot. *Sci. Rep.* **2018**, *8*, 3926. [CrossRef]
17. Hodgson, J.C.; Baylis, S.M.; Mott, R.; Herrod, A.; Clarke, R.H. Precision wildlife monitoring using unmanned aerial vehicles. *Sci. Rep.* **2016**, *6*, 22574. [CrossRef] [PubMed]
18. Chabot, D.; Craik, S.R.; Bird, D.M. Population census of a large common tern colony with a small unmanned aircraft. *PLoS ONE* **2015**, *10*, e0122588. [CrossRef] [PubMed]
19. McEvoy, J.F.; Hall, G.P.; McDonald, P.G. Evaluation of unmanned aerial vehicle shape, flight path and camera type for waterfowl surveys: Disturbance effects and species recognition. *PeerJ* **2016**, *4*, e1831. [CrossRef]
20. Dulava, S.; Bean, W.T.; Richmond, O.M.W. Applications of unmanned aircraft systems (UAS) for waterbird surveys. *Environ. Pract.* **2015**, *17*, 201–210. [CrossRef]

21. Drever, M.C.; Chabot, D.; O'Hara, P.D.; Thomas, J.D.; Breault, A.; Millikin, R.L. Evaluation of an unmanned rotorcraft to monitor wintering waterbirds and coastal habitats in British Columbia, Canada. *J. Unmanned Veh. Syst.* **2015**, *3*, 259–267. [CrossRef]

22. Junda, J.H.; Greene, E.; Zazelenchuk, D.; Bird, D.M. Nest defense behaviour of four raptor species (osprey, bald eagle, ferruginous hawk, and red-tailed hawk) to a novel aerial intruder—A small rotary-winged drone. *J. Unmanned Veh. Syst.* **2016**, *4*, 217–227. [CrossRef]

23. Weissensteiner, M.H.; Poelstra, J.W.; Wolf, J.B.W. Low-budget ready-to-fly unmanned aerial vehicles: An effective tool for evaluating the nesting status of canopy-breeding bird species. *J. Avian Biol.* **2015**, *46*, 425–430. [CrossRef]

24. Potapov, E.R.; Utekhina, I.G.; McGrady, M.J.; Rimlinger, D. Usage of UAV for surveying Steller's sea eagle nests. *Raptors Conserv.* **2013**, *27*, 253–260.

25. Kéry, M.; Schmidt, B.R. Imperfect detection and its consequences for monitoring for conservation. *Commun. Ecol.* **2008**, *9*, 207–216. [CrossRef]

26. Wilson, A.M.; Barr, J.; Zagorski, M. The feasibility of counting songbirds using unmanned aerial vehicles. *Auk* **2017**, *134*, 350–362. [CrossRef]

27. dos Santos, G.A.M.; Barnes, Z.; Lo, E.; Ritoper, B.; Nishizaki, L.; Tejeda, X.; Ke, A.; Lin, H.; Schurgers, C.; Lin, A.; et al. Small unmanned aerial vehicle system for wildlife radio collar tracking. In Proceedings of the 2014 IEEE 11th International Conference on Mobile Ad Hoc and Sensor Systems, Philadelphia, PA, USA, 28–30 October 2014; pp. 761–766. [CrossRef]

28. Stöcker, C.; Bennett, R.; Nex, F.; Gerke, M.; Zevenbergen, J. Review of the current state of UAV regulations. *Remote Sens.* **2017**, *9*, 459. [CrossRef]

29. Cracknell, A.P. UAVs: Regulations and law enforcement. *Int. J. Remote Sens.* **2017**, *38*, 3054–3067. [CrossRef]

drones

MDPI

Article

Drone Monitoring of Breeding Waterbird Populations: The Case of the Glossy Ibis

Isabel Afán [1],*, Manuel Máñez [2] and Ricardo Díaz-Delgado [1,2]

1   Remote Sensing and GIS Laboratory, Estación Biológica de Doñana (CSIC), 41092 Seville, Spain; rdiaz@ebd.csic.es
2   Natural Processes Monitoring Team, ICTS-RBD, Estación Biológica de Doñana, CSIC, 41092 Seville, Spain; mmanez@ebd.csic.es
*   Correspondence: isabelafan@ebd.csic.es; Tel.: +34-954-466700

Received: 1 November 2018; Accepted: 27 November 2018; Published: 1 December 2018

Abstract: Waterbird communities are potential indicators of ecological changes in threatened wetland ecosystems and consequently, a potential object of ecological monitoring programs. Waterbirds often breed in largely inaccessible colonies in flooded habitats, so unmanned aerial vehicle (UAV) surveys provide a robust method for estimating their breeding population size. Counts of breeding pairs might be carried out by manual and automated detection routines. In this study we surveyed the main breeding colony of Glossy ibis (*Plegadis falcinellus*) at the Doñana National Park. We obtained a high resolution image, in which the number and location of nests were determined manually through visual interpretation by an expert. We also suggest a standardized methodology for nest counts that would be repeatable across time for long-term monitoring censuses, through a supervised classification based primarily on the spectral properties of the image and a subsequent automatic size and form based count. Although manual and automatic count were largely similar in the total number of nests, accuracy between both methodologies was only 46.37%, with higher variability in shallow areas free of emergent vegetation than in areas dominated by tall macrophytes. We discuss the potential challenges for automatic counts in highly complex images.

Keywords: UAV; aerial survey; long-term monitoring; *Plegadis falcinellus*; bird censuses; supervised classification; image processing

1. Introduction

Ecological monitoring programs are fundamentally important for maintaining long-term series of data to evaluate the impact of anthropogenic activities and global change on protected areas [1–3]. Most conservation policies and management decisions rely primarily on information from monitoring programmes that focus on the ecosystem structures and processes and biodiversity variables as species populations [4–6]. One of the main goals of population monitoring programs is the study of species distribution changes over time. In this regard, it is crucial to develop consistent methods to monitor species distribution and abundance that will allow us to assess measures of population fluctuations [7].

Birds can be excellent barometers of environmental health [8], particularly in threatened habitats as wetlands, where both species trophic resources and breeding conditions largely depend on ecological and hydrological conditions. Accordingly, waterbird communities are extremely sensitive to changes in the availability of suitable wetland habitats, becoming indicator species to promote conservation awareness and actions [9]. However, as most species of waterbirds breed in largely inaccessible colonies within flooded habitats, it is difficult to carry out surveys of colonial wetland birds in situ. Many of these bird species breed densely aggregated in inaccessible places with no possibility of performing ground surveys without causing serious disturbances to the breeding individuals [10]. Furthermore, the breeding places often contain several species breeding in sympatry. Consequently, aerial surveys have become one of the most common methods for censusing colonies of waterbirds [11–14]. One of the main constraints of this methodology is the limited spatial resolution resulting from aircrafts flights, often not suited to local-scale ecological investigations [15]. Therefore, the application of UAVs has been a qualitative leap in bird monitoring, enabling rapid, low disturbance surveys of inaccessible areas while delivering repeatable images with a fine spatial resolution [16–19].

Despite the fine spatial resolution data that UAVs can deliver, there is some bias inherent in this methodology that could hamper the extraction of accurate population data of densely colonial aggregations of birds: the absence of ground-truth data in inaccessible habitats, the difficulty of isolating and quantifying individuals or nests because of their cryptic nature or poor visibility through the canopy [11] and the potential negative interactions between bird and drones [20,21]. Furthermore, the quality of sensors is rapidly increasing, resulting in progressively very large images which are difficult to treat with standard image processing software. Based on historical aerial surveys of waterbirds, counting error has proved to be one of the most serious problems for aerial surveys [14]. Individuals or nest counts have primarily relied on manual interpretation of the images produced. This can be time-consuming if imagery is collected over large areas and dense congregations of birds [22]. Thus, the use of both manual and automated detection routines for features counting and the development of automatic tools to detect large numbers of birds automatically from images has become a necessary subject matter for current research [21–23].

This study was developed on Doñana National Park (SW Spain), one of the largest protected wetlands in Europe with a typical Mediterranean climate. Doñana holds a large extent of temporary marshes (340 km^2). Wetland flooding patterns fluctuate interannual and seasonally, with natural climatic variability or anthropic modifications, modifying substantially the suitable area for breeding species [24]. Doñana is also recognized as a Long-Term Socio-Ecological Research (LTSER) platform integrated in the LTER-Europe network [25] by applying harmonized protocols for long-term socio-ecological research. We particularly focus on observations of a colonial species, the Glossy ibis *Plegadis falcinellus*, a species nesting in Doñana [26] where it has undergone an explosive population expansion from seven pairs in 1996 to more than 10,000 pairs by 2017 [27–30]. Glossy ibis is considered as least concern status by the International Union for the Conservation of Nature (IUCN) by its extremely large distribution range [31], but is cited as a conservation concern in the Spanish Red Data Book [32]. However, the species is threatened by wetland habitat degradation and loss [29,33], so their surveillance has a high priority within the long-term monitoring programs of Doñana National Park. Furthermore, the number of breeding pairs has increased dramatically in recent years in the FAO colony, with a total population in Doñana wetlands exceeding 10,000 pairs in 2017 [30]. This, combined with the fact that the species breeds in marshes, with nests positioned less than one meter above water in tall dense stands of emergent vegetation [29], makes ground censuses totally out of reach. The cryptic nature and dark colours of this species represent also a challenge for their discrimination by way of aerial censuses.

The purpose of this study was to evaluate the capabilities of UAVs to perform aerial censuses of densely populated wetland colonies of waterbirds, with the aim of integrating unmanned aerial flights in the long-term monitoring bird populations censuses. Our particular objective was to estimate the number of breeding pairs of Glossy ibis in the study area, comparing the capability of manual and automatic count procedures based on image-analysis techniques that use the spectral, size and form characteristics of the target species for an automatic recognition [14,23]. The final objective is to evaluate whether UAV surveys can be integrated as an efficient and standardized tool for long-term monitoring of colonial breeding waterbirds in Doñana National Park.

2. Materials and Methods

2.1. Species and Study Area

The first breeding report for Glossy Ibis at Doñana wetlands dates back to 1770 [34], but the species was pushed to extinction by hunting and the pillaging of the eggs during the early years of the twentieth century. It was not until 1996 that the species became re-established and bred regularly in Doñana. This natural area contained 90% of the Spanish breeding population of Glossy ibis in 2001 [29]. At present, "Lucio de la FAO" hosts the main Glossy ibis colony at Doñana National Park, a large mixed colony where it breeds in association with other species, regularly Ardeidae (Figure 1a).

Figure 1. (a) Glossy ibis (*Plegadis falcinellus*) individual. (b) Front view of the Glossy ibis colony, densely covered by the macrophyte *Typha*, which makes it extremely difficult the ground census of the colony. Photos: Carlos Gutiérrez-Expósito (a), and Manuel Máñez (b).

The "Lucio de la FAO" is a system of three interconnected semi-artificial ponds covering a total surface area of about 50 ha which is flooded by both direct precipitation and groundwater pumped from the underlying aquifer. The vegetation cover varies with locally dense stands of *Typha* (mostly used for nesting, Figure 1b), areas dominated by tall *Tamarix* scrub and shallow areas free of emergent vegetation [26].

2.2. Drone Survey Method

One aerial survey was conducted over the main breeding colony of Glossy ibis in Doñana National Park (37°4'23"N, 6°2'46"W, Figure 2a,b) on 14 May 2018 between 10:00 and 10:33 hours (local time). The survey was done using a Phantom 4 Pro quadcopter (DJI Innovations, Shenzhen, China). This UAV was equipped with a compact 4K digital camera of 20 megapixels (24 mm lens). The exposure time was set automatically using "speed priority" mode. The Phantom UAV was remotely controlled with a handheld unit keeping the legal maximum distance of 500 m from the operator.

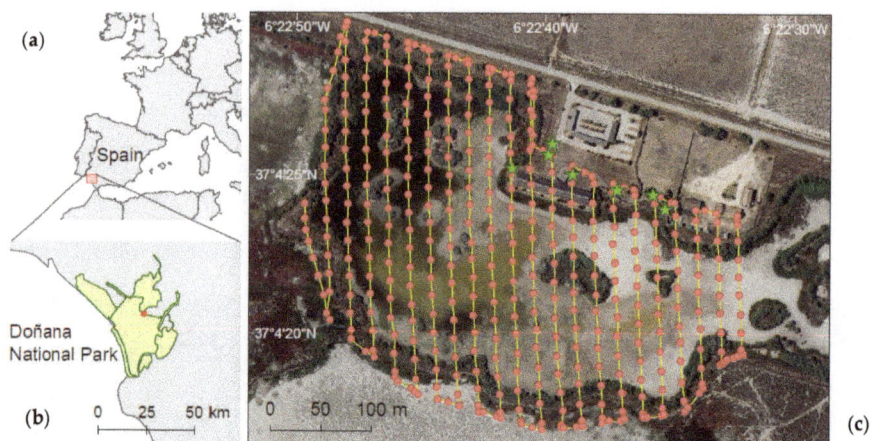

Figure 2. Location of Doñana National Park, SW Spain (**a,b**). Red dots indicate the situation of the Glossy ibis colony. (**c**) Automatic flight path covered by the UAV to collect image of the Glossy ibis colony in "Lucio de la FAO". The UAV followed 22 predefined transect lines (yellow lines, 6292.2 m) with 80% front overlap and 70% side overlap. Green stars show ground control points (GCPs) established to increase the absolute global accuracy of drone image.

The UAV followed an automatic flight programmed beforehand with 22 transect lines covering an area of 16.1 ha with 80% front overlap and 70% side overlap (see Figure 2c). The flight settings were defined using Pix4DCapture mobile app (Pix4D SA). The drone was flown at an altitude of 50 m above ground elevation and at a ground speed of 12 km h^{-1} (Figure 3a,b). A total of 448 images were taken. The ground sample distance (GSD), being the distance between adjacent pixel centres on the ground was 0.0132 m. Environmental conditions provide optimal flight conditions: high visibility (>1000 m), no cloud cover and wind speeds <4 m s^{-1}. The total duration of the census was ca. 1 h. No adverse behaviours from birds against the UAV or disturbance reactions were observed.

Flight was operated during the late hatching period of the Glossy ibis, to assure that only one individual from the breeding pair was lying on the nest, while the other fed in the nearby foraging marshland areas. Both sexes of Glossy ibis individuals alternate to incubate the eggs and feed the chicks [35], and the exchange takes place early in, and at the end of the day, with no interference during the flight time.

The pictures were mosaicked with the Pix4Dmapper photogrammetric software (Pix4D SA, Switzerland) to obtain a single multiband RGB image of the colony. The Pix4D computing consists of three steps: initial processing, point cloud densification and DSM and orthomosaic generation, which are carried out automatically [36]. The final spatial resolution of the orthomosaic was 0.0132 m. Only seven ground control points (GCPs) could be placed using conspicuous elements in the accessible area (see in Figure 2 the paths surrounding the main building) providing a mean RMS (root mean square error) of 0.06 m.

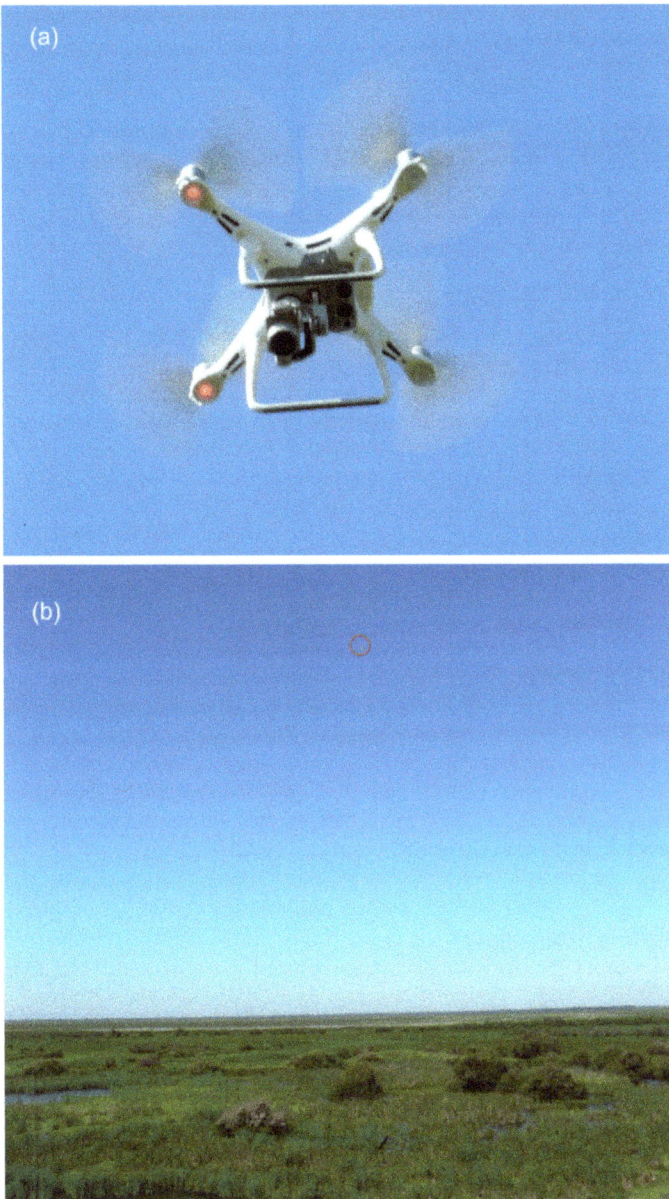

Figure 3. (a) Drone used to survey the Glossy ibis colony, Phantom 4 Pro quadcopter. (b) Phantom 4 surveying the colony at 50 m of altitude. Birds show no disturbance. Photos: Manuel Máñez.

2.3. Visual Identification of Breeding Birds

Confronted with an inability to collect ground-truth data for several of the nesting birds in the colony, we performed a manual and visual identification of the breeding birds in the orthomosaic. An experienced ornithologist identified birds sitting on nests 'by eye'. A point vector file with nest positions was created while onscreen digitizing at 1:30 scale. Other sympatric breeding species, whenever visible, were also geolocated in the image. Visual identification was performed in ArcGIS 10.5 (ESRI, Redlands, California, U.S.A.).

2.4. Automatic Identification of Breeding Birds

The automatic detection of breeding individuals was accomplished in two steps: first we performed a supervised classification based primarily on the spectral properties of the image with the aim of simplifying the original complexity of the entire scene and enhancing the subsequent automatic detection of Glossy ibis individuals. In a second stage, we carried out an automatic count using image processing software where size and form were introduced to the recognition process.

2.4.1. Supervised Classification

A supervised classification segments the spectral domain of an image and assigns every pixel a probability of belonging to one of the classes of interest, based on the spectral information of homogeneous training areas for the different thematic classes being considered [37]. We used a Random Forest supervised classification to classify the original image into three classes. The Random Forest classifier consists of a combination of decision trees from randomly selected subset of a training set. The trees are created by drawing a subset of training samples through replacement: two-thirds of the data are sampled for training and the remaining third of the data are excluded for validation. The final classification decision is taken by averaging the class assignment probabilities calculated by all produced trees. A new unlabelled data input is thus evaluated against all decision trees created in the ensemble and each tree votes for a class membership. Pixels are classified by taking the most popular voted class from all the tree predictors in the forest [38,39]. Random Forests are widely popular because of their ability to classify large amounts of data with high accuracy and preventing overfitting [39]. We used breeding birds sitting on their nests from the orthomosaic as training areas for the class "ibis". Nest structures, typically completely covered by guano, represent the second class. The third class evaluated comprised areas covered by a homogenous film of *Lemna*. The remaining more complex vegetation and other species (i.e., Purple heron) were not considered because those classes shares spectral properties with the Glossy ibis and their inclusion considerably reduced the accuracy of classification. A total of 33,977 pixels were selected for the training features, with a similar number of pixels for the three classes (class balanced). A total of 100 trees were used. The resulting image was reclassified to maintain only the category "ibis".

All analyses were conducted in R [40]. The packages rgdal [41], raster [42], caret [43], and randomForest [44] were also used.

2.4.2. Image Processing

Image analyses were performed with the freely available Fiji distribution of ImageJ software (developed at the US National Institutes of Health) [45,46] and the "Analyse Particles" function which counts objects in binary or thresholded images. Supervised classification and subsequent reclassification through a binary image was used as input for the image analysis. This step avoids the need to select a threshold value which will prevent objects above a certain pixel value being included in the automatic count total.

One hundred Glossy ibis individuals randomly distributed in the study area were digitalized with a 600× zoom. Size and form (circularity) measures were extracted to determine the upper and lower threshold limiting size and form-based selection of particles to be counted. The resulting grid classified image showing the detected particles was vectorized in ArcGIS 10.2 (ESRI, Redlands, California, U.S.A.)., applying first a low pass filter and a majority filter to soften the limits of the areas classified as ibis individuals. Micropolygons with a surface area of less than 0.007 m^2, which had split from complex shaped polygons during the vectorization process, were removed.

2.5. Validation

To test whether the automatic method was reliable, accuracy (proportion of nests mapped correctly), commission errors (proportion of nests counted automatically but left out in the manual process) and omission errors (proportion of nests omitted in the automatic procedure) were determined by spatial comparison of georeferenced nests obtained from the automatic process to those visually determined over the image by an expert. Validation was performed on the total colony and on six square sample plots of 329 m², selected over the main vegetation types where nests were settled, as derived from visual inspection of the orthomosaic obtained: three sample plots in the flooded areas densely covered by macrophytes (mainly *Typha*, number of nests ranged 94–228) and three in the free areas of emergent vegetation but covered with a layer of aquatic plants (*Lemna* spp, number of nests ranged 126–161).

Spatial overlap was conducted between the individuals obtained from the vectorization of the automatic classified image and a buffer area of 0.15 m of radius around the nests detected manually. A visual inspection was also performed to assure that all identified nests were inside their corresponding overlap area.

Validation plots were also used for estimating Glossy ibis nest density and nest distance in each vegetation class from the manual georeferenced nests.

3. Results

The orthomosaic obtained from the aerial survey showed an area of 16 ha covering the complete surface of the Lucio de la FAO where Glossy ibis breeds (Figure 4). The Glossy ibis colony was based in a continuous area of 3.2 ha with two major classes of vegetation cover: flooded areas covered by macrophytes (mainly *Typha*) and free areas of emergent vegetation but covered with aquatic plants (*Lemna* spp). Regardless the vegetation type cover around the nests, *Typha* invariably constitute the predominant nesting material (Figure 4a). No nests were found over *Tamarix* shrubs.

Figure 4. Orthomosaic image of high-resolution of the Glossy ibis colony in "Lucio de la FAO". Image was mosaicked with Pix4D mapper Pro (Pix4D, Switzerland) after acquiring 448 photos through automatic flight of Phantom 4 Pro quadcopter (DJI Innovations, Shenzhen, China) on 14 May 2018. Squares show validation areas to test the accuracy of the automatic nest counting method to the visually nest count by an expert. Blue squares show densely macrophytes area (*Typha* spp) whereas red squares show free areas of emergent vegetation covered by *Lemna* aquatic plant. White line delineates the Glossy ibis nesting area.

3.1. Manual Counting

Visual identification of lying birds on nest identified a total of 7134 individuals of Glossy ibis together with 439 individuals of Purple heron (*Ardea purpurea*). Glossy ibis nests measured 0.57 ± 0.1 m (mean $\pm$ SD) in diameter (n = 50), similar to those of Purple heron (0.70 ± 0.24, n = 50). Glossy ibis nests were densely distributed throughout the colony (0.22 nests m^{-2}), with similar density in areas covered by different vegetation classes, as *Typha* or *Lemna* areas (0.4 nests m^{-2}, Table 1). Minimum proximity distance within nests was lower than 0.5 m in all areas, with maximum proximity distances up to 7 m (Table 1). Purple heron nests were not uniformly distributed into the Glossy ibis breeding area, with a preference for areas with greater vegetation cover of *Typha*.

Table 1. Glossy ibis nest density, minimum nest distances (mean $\pm$ SD and range) obtained from manual georeferencing of nests, in the total colony and inside the three validation areas selected for each vegetation class. The number of total Glossy ibis nests and range number of nests considered for each vegetation class is also shown.

Area	Nest Density (Nests m^{-2})	Nest Distance (m)	Nest Distance (Range, m)	Number of Nests
Total colony	0.22	0.88 ± 0.48	0.26–7.84	7134
Typha spp.	0.41 ± 0.22	0.82 ± 0.36	0.32–3.85	94–228
Lemna spp.	0.43 ± 0.05	0.79 ± 0.42	0.33–3.35	128–161

3.2. Automatic Counting

The automatic bird count showed a total of 8,060 nesting birds, but was unable to distinguish between the two species breeding in the colony (Figure 5). Total accuracy between nest counts manually and nests counted by the automatic procedure was of 43.65%. Accuracy was similar in areas with dense canopy covered by macrophytes (*Typha*) (50.37%, Table 2), than in open areas only covered by *Lemna* (49.01%, Table 2). Nevertheless, variability between validation plots was much higher in *Lemna* than in *Typha* areas (Table 2).

Table 2. Accuracy (total agrees with the percent of total nests detected), commission, and omission errors obtained between manual count and automatic nest counting procedure, in the total colony and in average and SD for the three validation plots selected for each vegetation classes. The number of total Glossy ibis nests and range of nests enclosed in the three validation areas considered for each vegetation class is also shown.

Area	Accuracy (%)	Commission Error (%)	Omission Error (%)	Number of Nests
Total colony	46.37	66.61	53.63	7134
Typha spp.	50.37 ± 2.19	55.06 ± 27.79	49.63 ± 2.19	94–228
Lemna spp.	49.01 ± 19.22	22.64 ± 7.45	50.99 ± 19.22	128–161

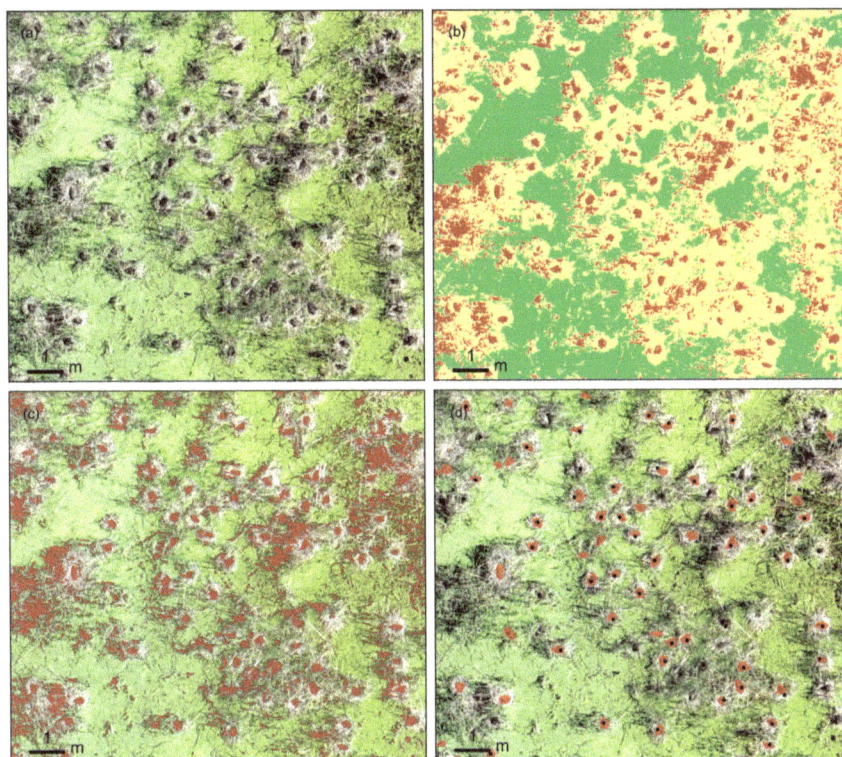

Figure 5. Automatic Glossy ibis nests count process. (**a**) Overview of the Glossy ibis nests from the orthomosaic obtained by the UAV. (**b**) Result of the Random Forest supervised classification. Pixels were classified into three classes: Glossy ibis (red), nest structures (yellow) and *Lemna* cover (green). (**c**) Two-bits image, result of the reclassification of the supervised classified image to maintain only ibis class. (**d**) Glossy ibis laying individuals automatically obtained by the image processing software (ImageJ), size and form based. Dots show identified nests by visual inspection (black dots indicate Glossy ibis nests and yellow dot indicates Purple heron nest).

4. Discussion

The survey conducted by means of UAV technology, allowed us to obtain a high-resolution image of the main Glossy ibis colony in the Doñana National Park, settled in a flooded inaccessible area. From this orthomosaic, we determined the nests established, their spatial location and the presence of another species (Purple heron), albeit in a minority, breeding in sympatry with the Glossy ibis. The spatial resolution accomplished by the flight allowed a good by-eye estimation of the breeding population, since nest structures and Glossy ibis individuals were able to be identified in the orthomosaic. Despite this, the automatic count was less accurate than the manual count, with the automatic count consistently failing to detect the 50% of the birds in the image.

The census of Glossy ibis breeding in Lucio de la FAO was in fact a challenging task: this colony hosts a high density of individuals, nests are constructed inside macrophytic vegetation up to more than one meter in height, and the target species has a dark-coloured plumage, which makes it difficult to stand out from the background. Indeed, aerial censuses of this species had produced very poor outcomes, detecting only 17% of the nests in freshwater marshes [11]. Therefore, the accuracy rates in waterbird colony manual counts have clearly improved with the use of UAVs technology, allowing near 100% of nests recognition with image resolutions under 0.01 m, and ideally when reached 0.005 m [10]. Dulava et al. [47] recommended an image resolution of ~0.005 m/pixel to enable correct species identification [48]. Object-oriented approaches also recommend very high resolution images (0.01 m/pixel or finer) [23]. With a spatial resolution of 0.01319 m/pixel obtained in our trial, performance shortcomings in automatic counts in this work can be likely attributed in part, to an insufficient image resolution. Furthermore, the automatic count omitted those individuals not well isolated from their background. Shaded vegetation and dark areas of water were assigned to the same group as dark Glossy ibis individuals in the course of the supervised classification process, causing commission errors during the counting process. On the other hand, white guano areas laid on the nest material, which were easily recognized throughout the classification, allowed a good isolation of breeding individuals. However, nests with a reduced white guano envelope were more likely to be omitted during the automatic count process. The automatic count performed similar at nest detection in canopied macrophytes areas (*Typha* spp) than in open waters covered by *Lemna*. However, variability in detection was much higher in *Lemna* areas. Low detection may be related to patches of dark water within the *Lemna* beds could be mistaken for the 'Ibis class' if they were within the size limit determined for Glossy Ibis particles.

While an improvement of the spatial and radiometric resolution of the image may provide a better outcome, we propose here a methodology that allows a standardized automatic nesting count, easily implementable and repeatable across time for long-term monitoring censuses. Reducing the complexity of images for extracting information usually relies on a thresholding step, which transforms the original image into a binary one, in which a cut-off value determines the level above and below which the pixels will be selected for being measured [49]. Consequently, this is one of the most critical and at the same time subjective steps in the image processing. Thresholding has proved to be highly effective in cases where target features are spatially separated and where they do not share the same spectral range as other image elements in the background [50–52]. However, in situations without high bird contrast, classifications can be considered as an alternative approach to isolate birds from background features [52–54]. With regard to the complexity of the Glossy ibis colony image, we have performed a supervised classification for simplifying the spectral classes of the original image and reduce the file size. Both processes were needed to enhance the capabilities of image processing software (ImageJ). In addition, size and shape filtering, particularly roundness, which has been proved to be effective detecting several bird species [52,53] was subsequently used.

The results derived from this study demonstrate the capability of UAV censuses for detailed monitoring of Glossy ibis (and, therefore, other waterbirds breeding in similar habitats), and their applicability for obtaining long term comparable breeding population trends. For many waterbird populations, aerial surveys are proving the only way to collect data over large areas at a relatively low cost [14]. This case study reveals the benefits and limitations of the applicability of UAVs for waterbirds monitoring: drones are capable of collecting high-resolution spatial data in difficult access areas, with non-significant disturbance to the breeding birds, and with an affordable cost depending on the surface to be covered [16]. In this particular case, the colony could be flown over in a short time (one hour), which was helped by the proximity of access routes, which results in a suitable compromise between cost and results obtained. The extraordinary growth in recent years of the breeding population of this species means that any other methodology for obtaining breeding data is unfeasible. The Glossy ibis colony needs a low altitude flightfor providing images of sufficiently high resolution to accurately isolate individual nests in the image, but without causing disturbance to the breeding birds and their chicks. Indeed, the resolution accomplished to date allows a manual approach to counting nests, but is still unsatisfactory for an automatic estimate. As UAV capabilities will increase, in terms of flight endurance and sensor resolution, a larger amount of high-resolution data will be collected, and automatic processing will become a major need [55]. Adjusting automatic procedures is essential to incorporate UAV data into long-term monitoring programs, since this will reduce bias due to visual interpretation and provide comparable data over time.

Author Contributions: Conceptualization, I.A., M.M. and R.D.-D.; Data curation, M.M. and R.D.-D.; Formal analysis, I.A., M.M. and R.D.-D.; Methodology, I.A., M.M. and R.D.-D.; Writing—original draft, I.A.; Writing—review & editing, M.M. and R.D.-D.

Funding: Consejería de Medio Ambiente y Ordenación del Territorio of Junta de Andalucía and the ICTS program by the Spanish Ministry of Science and Technology provide funding for the Long Term Monitoring Program of Doñana Natural Space including drone flights. Waterbird surveys have been funded through a contract with Environment and Water Agency (AMAYA), of the Regional Environment Authority of Andalusia.

Acknowledgments: We thank to the Natural Processes Monitoring Team (Birds) from Estación Biológica de Doñana (currently formed by Luis García, José Luis del Valle, José Luis Arroyo, Rubén Rodríguez and Antonio Martínez) for all the effort to the collect of long monitoring data on Glossy ibis. We are very grateful to Clive Hurford for their constructive review and David Aragonés and Francisco Ramírez for helpful comments.

Conflicts of Interest: The authors declare no conflict of interest.

Statements: This study complied with Spanish legislation (Law 48/1960 and Reales Decretos 1036/2017, 552/2014, and 57/2002) defining regulation for civil use of Remotely Piloted Aircraft Systems (RPAS). EBD-CSIC is a legal listed operator by AESA since 2017 and as CSIC since 2015. Declared activities include research and development (https://www.seguridadaerea.gob.es/media/4305572/listado_operadores.pdf page 181). The radio frequency used for UAS radio control was 35 MHz, which is the frequency authorized for radio control for leisure applications in Europe under directive 1999/05/EC. The project had the permission of the National Park authorities on nature conservation.

References

1. Haase, P.; Tonkin, J.D.; Stoll, S.; Burkhard, B.; Frenzel, M.; Geijzendorffer, I.R.; Hauser, C.; Klotz, S.; Kuhn, I.; McDowell, W.H.; et al. The next generation of site-based long-term ecological monitoring: Linking essential biodiversity variables and ecosystem integrity. *Sci. Total Environ.* **2018**, *613–614*, 1376–1384. [CrossRef] [PubMed]

2. Parmesan, C.; Burrows, M.T.; Duarte, C.M.; Poloczanska, E.S.; Richardson, A.J.; Schoeman, D.S.; Singer, M.C. Beyond climate change attribution in conservation and ecological research. *Ecol. Lett.* **2013**, *16*, 58–71. [CrossRef]

3. Magurran, A.E.; Baillie, S.R.; Buckland, S.T.; Dick, J.M.; Elston, D.A.; Scott, E.M.; Smith, R.I.; Somerfield, P.J.; Watt, A.D. Long-term datasets in biodiversity research and monitoring: assessing change in ecological communities through time. *Trends Ecol. Evol.* **2010**, *25*, 574–582. [CrossRef] [PubMed]

4. Nager, R.G.; Hafner, H.; Johnson, A.R.; Cézilly, F. Environmental Impacts on Wetland Birds: Long-Term Monitoring Programmes in the Camargue, France. *Ardea* **2010**, *98*, 309–318. [CrossRef]

5. Grumbine, R.E. What is ecosystem management? *Conserv. Biol.* **1994**, *8*, 27–38. [CrossRef]

6. Lambeck, R.J. Focal species: A multi-species umbrella for nature conservation. *Conserv. Biol.* **1997**, *11*, 849–856. [CrossRef]

7. Vos, P.; Meelis, E.; Ter Keurs, W.J. A framework for the design of ecological monitoring programs as a tool for environmental and nature management. *Environ. Monit. Assess.* **2000**, *61*, 317–344. [CrossRef]

8. Sutherland, W.J.; Newton, I.; Green, R. *Bird Ecology and Conservation: A Handbook of Techniques*; Oxford University Press: Oxford, UK, 2004; Volume 1.

9. Ramírez, F.; Rodríguez, C.; Seoane, J.; Figuerola, J.; Bustamante, J. How will climate change affect endangered Mediterranean waterbirds? *PLOS ONE* **2018**, *13*, e0192702. [CrossRef]

10. Bako, G.; Tolnai, M.; Takacs, A. Introduction and testing of a monitoring and colony-mapping method for waterbird populations that uses high-speed and ultra-detailed aerial remote sensing. *Sensors* **2014**, *14*, 12828–12846. [CrossRef]

11. Frederick, P.C.; Towles, T.; Sawicki, R.J.; Bancroft, G.T. Comparison of aerial and ground techniques for discovery and census of wading bird (Ciconiiformes) nesting colonies. *The Condor* **1996**, *98*, 837–841. [CrossRef]

12. Kushlan, J.A. Effects of helicopter censuses on wading bird colonies. *J. Wildlife Manage.* **1979**, *43*, 756–760. [CrossRef]

13. Díaz-Delgado, R. An integrated monitoring programme for Doñana Natural Space: The set-up and implementation. In *Conservation Monitoring in Freshwater Habitats: A Practical Guide and Case Studies*; Hurford, C., Schneider, M., Cowx, I., Eds.; Springer Netherlands: Dordrecht, The Netherlands, 2010; pp. 325–337.

14. Kingsford, R.T.; Porter, J.L. Monitoring waterbird populations with aerial surveys what have we learnt? *Wildl. Res.* **2009**, *36*, 29–40. [CrossRef]
15. Anderson, K.; Gaston, K.J. Lightweight unmanned aerial vehicles will revolutionize spatial ecology. *Front. Ecol. Environ.* **2013**, *11*, 138–146. [CrossRef]
16. Sardà-Palomera, F.; Bota, G.; Padilla, N.; Brotons, L.; Sardà, F. Unmanned aircraft systems to unravel spatial and temporal factors affecting dynamics of colony formation and nesting success in birds. *J. Avian Biol.* **2017**, *48*, 1273–1280. [CrossRef]
17. Hodgson, J.C.; Baylis, S.M.; Mott, R.; Herrod, A.; Clarke, R.H. Precision wildlife monitoring using unmanned aerial vehicles. *Sci. Rep.* **2016**, *6*, 22574. [CrossRef] [PubMed]
18. Brisson-Curadeau, E.; Bird, D.; Burke, C.; Fifield, D.A.; Pace, P.; Sherley, R.B.; Elliott, K.H. Seabird species vary in behavioural response to drone census. *Sci. Rep.* **2017**, *7*, 17884. [CrossRef] [PubMed]
19. Han, Y.-G.; Yoo, S.H.; Kwon, O. Possibility of applying unmanned aerial vehicle (UAV) and mapping software for the monitoring of waterbirds and their habitats. *J. Ecol. Environ.* **2017**, *41*. [CrossRef]
20. Fuller, A.R.; McChesney, G.J.; Golightly, R.T. Aircraft disturbance to Common Murres (*Uria aalge*) at a breeding colony in Central California, USA. *Waterbirds* **2018**, *41*, 257–267. [CrossRef]
21. Lyons, M.; Brandis, K.; Callaghan, C.; McCann, J.; Mills, C.; Ryall, S.; Kingsford, R. Bird interactions with drones, from individuals to large colonies. *Australian Field Ornithol.* **2018**, *35*. [CrossRef]
22. Descamps, S.; Béchet, A.; Descombes, X.; Arnaud, A.; Zerubia, J. An automatic counter for aerial images of aggregations of large birds. *Bird Study* **2011**, *58*, 302–308. [CrossRef]
23. Grenzdörffer, G.J. UAS-based automatic bird count of a common gull colony. *Int. Arch. Photogramm. Remote Sens. Spat. Inf. Sci.* **2013**, *1*, 169–174. [CrossRef]
24. Díaz-Delgado, R.; Aragonés, D.; Afán, I.; Bustamante, J. Long-term monitoring of the flooding regime and hydroperiod of Doñana marshes with Landsat time series (1974–2014). *Remote Sens.* **2016**, *8*. [CrossRef]
25. Haberl, H.; Gaube, V.; Díaz-Delgado, R.; Krauze, K.; Neuner, A.; Peterseil, J.; Plutzar, C.; Singh, S.J.; Vadineanu, A. Towards an integrated model of socioeconomic biodiversity drivers, pressures and impacts. A feasibility study based on three European long-term socio-ecological research platforms. *Ecol. Econ.* **2009**, *68*, 1797–1812. [CrossRef]
26. Santoro, S.; Máñez, M.; Green, A.J.; Figuerola, J. Formation and growth of a heronry in a managed wetland in Doñana, southwest Spain. *Bird Study* **2010**, *57*, 515–524. [CrossRef]
27. Santoro, S.; Green, A.J.; Speakman, J.R.; Figuerola, J. Facultative and non-facultative sex ratio adjustments in a dimorphic bird species. *Oikos* **2015**, *124*, 1215–1224. [CrossRef]
28. Ramo, C.; Aguilera, E.; Figuerola, J.; Máñez, M.; Green, A.J. Long-term population trends of colonial wading birds breeding in Doñana (SW Spain) in relation to environmental and anthropogenic factors. *Ardeola* **2013**, *60*, 305–326. [CrossRef]
29. Figuerola, J.; Máñez, M.; Ibáñez, F.; García, L.; Garrido, H. Morito Común, *Plegadis falcinellus*. In *Atlas de las Aves Reproductoras de España*; Martí, R., del Moral, J.C., Eds.; Dirección General de Conservación de la Naturaleza-SEO/BirdLife: Madrid, Spain, 2004; pp. 124–125.
30. Máñez, M.; García, L.; Arroyo, J.L.; Del Valle, J.L.; Rodríguez, R.; Martínez, M.; Chico, A. Twenty-two years of monitoring of the Glossy Ibis (*Plegadis falcinellus*) in Doñana. In Proceedings of the First International Workshop on Glossy Ibis, Doñana, Spain, 27–29 November 2017.
31. BirdLife International. Plegadis falcinellus. The IUCN Red List of Threatened Species. 2016. Available online: http://dx.doi.org/10.2305/IUCN.UK.2016-3.RLTS.T22697422A86436401.en (accessed on 19 November 2018).
32. Madroño, A.; González, G.G.; Atienza, J.C. *Libro rojo de las aves de España*; Dirección General para la Biodiversidad-SEO/BirdLife: Madrid, Spain, 2004.
33. Del Hoyo, J.; Elliot, A.; Sargatal, J. *Handbook of the Birds of the World*; Lynx Editions: Barcelona, Spain, 1992.
34. Valverde, J.A. *Vertebrados de las marismas del Guadalquivir (introducción al estudio ecológico)*; Archivos del Instituto de Aclimatación: Almería, Spain, 1960; Vol. IX.
35. Cramp, S.; Simmons, K.E.L.; Perrins, C.M. *The Birds of the Western Palearctic*; Oxford University Press: Oxford, UK, 1977; Vol. I.
36. Ivosevic, B.; Han, Y.-G.; Kwon, O. Calculating coniferous tree coverage using unmanned aerial vehicle photogrammetry. *J. Ecol. Env.* **2017**, *41*. [CrossRef]

37. Richards, J.A.; Richards, J.A. *Remote Sensing Digital Image Analysis*; Springer: Berlin/Heidelberg, Germany, 1999; Volume 3.
38. Pal, M. Random forest classifier for remote sensing classification. *Int. J. Remote Sens.* **2007**, *26*, 217–222. [CrossRef]
39. Belgiu, M.; Drăguţ, L. Random forest in remote sensing: A review of applications and future directions. *ISPRS J. Photogramm. Remote Sens.* **2016**, *114*, 24–31. [CrossRef]
40. R Core Team. *R: A Language and Environment for Statistical Computing*; R Foundation for Statistical Computing: Vienna, Austria, 2017; Available online: http://www.R-project.org/ (accessed on 22 October 2018).
41. Bivand, R.; Keitt, T.K.; Rowlingson, B. *rgdal: Bindings for the 'Geospatial' Data Abstraction Library*, R package version 1.2-16; 2018. Available online: https://cran.r-project.org/web/packages/rgdal/index.html (accessed on 22 October 2018).
42. Hijmans, R.J. *raster: Geographic Data Analysis and Modeling*, R package version 2.6-7; 2017. Available online: https://cran.r-project.org/web/packages/raster/index.html (accessed on 22 October 2018).
43. Kuhn, M.; Wing, J.; Weston, S.; Williams, A.; Keefer, C. *caret: Classification and Regression Training*, R package version 6.0-80; 2012. Available online: https://cran.r-project.org/web/packages/caret/index.html (accessed on 22 October 2018).
44. Liaw, A.; Wiener, M. Classification and Regression by randomForest. *R News* **2002**, *2*, 18–22.
45. Schindelin, J.; Arganda-Carreras, I.; Frise, E.; Kaynig, V.; Longair, M.; Pietzsch, T.; Preibisch, S.; Rueden, C.; Saalfeld, S.; Schmid, B. Fiji: an open-source platform for biological-image analysis. *Nat. Meth.* **2012**, *9*, 676. [CrossRef] [PubMed]
46. Rueden, C.T.; Schindelin, J.; Hiner, M.C.; DeZonia, B.E.; Walter, A.E.; Arena, E.T.; Eliceiri, K.W. ImageJ2: ImageJ for the next generation of scientific image data. *BMC Bioinform.* **2017**, *18*, 529. [CrossRef] [PubMed]
47. Dulava, S.; Bean, W.T.; Richmond, O.M. Environmental reviews and case studies: Applications of unmanned aircraft systems (UAS) for waterbird surveys. *Environ. Pract.* **2015**, *17*, 201–210. [CrossRef]
48. Barr, J.R.; Green, M.C.; DeMaso, S.J.; Hardy, T.B. Detectability and visibility biases associated with using a consumer-grade unmanned aircraft to survey nesting colonial waterbirds. *J. Field Ornithol.* **2018**, *89*, 242–257. [CrossRef]
49. Mallard, F.; Le Bourlot, V.; Tully, T. An automated image analysis system to measure and count organisms in laboratory microcosms. *PLOS ONE* **2013**, *8*, e64387. [CrossRef] [PubMed]
50. Trathan, P.N. Image analysis of color aerial photography to estimate penguin population size. *Wildlife Soc. B.* **2004**, *32*, 332–343. [CrossRef]
51. Hurford, C. Improving the accuracy of bird counts using manual and automated counts in ImageJ: An open-source image processing program. In *The Roles of Remote Sensing in Nature Conservation: A Practical Guide And Case Studies*; Díaz-Delgado, R., Lucas, R., Hurford, C., Eds.; Springer International Publishing: Cham, Switzerland, 2017; pp. 249–276. [CrossRef]
52. Chabot, D.; Francis, C.M. Computer-automated bird detection and counts in high-resolution aerial images: A review. *J. Field Ornithol.* **2016**, *87*, 343–359. [CrossRef]
53. Liu, C.-C.; Chen, Y.-H.; Wen, H.-L. Supporting the annual international black-faced spoonbill census with a low-cost unmanned aerial vehicle. *Ecol. Inform.* **2015**, *30*, 170–178. [CrossRef]
54. Díaz-Delgado, R.; Máñez, M.; Martínez, A.; Canal, D.; Ferrer, M.; Aragonés, D. Using UAVs to map aquatic bird colonies. In *The Roles of Remote Sensing in Nature Conservation*; Springer: Cham, Switzerland, 2017; pp. 277–291.
55. Linchant, J.; Lisein, J.; Semeki, J.; Lejeune, P.; Vermeulen, C. Are unmanned aircraft systems (UASs) the future of wildlife monitoring? A review of accomplishments and challenges. *Mammal. Rev.* **2015**, *45*, 239–252. [CrossRef]

![drones](drones logo) *drones*

MDPI

Article

Assessment of Chimpanzee Nest Detectability in Drone-Acquired Images

Noémie Bonnin [1,*], Alexander C. Van Andel [2], Jeffrey T. Kerby [3], Alex K. Piel [1], Lilian Pintea [4] and Serge A. Wich [1,5]

[1]   School of Natural Sciences and Psychology, Liverpool John Moores University, Liverpool L3 3AF, UK;
      A.K.Piel@ljmu.ac.uk (A.K.P.); S.A.Wich@ljmu.ac.uk (S.A.W.)
[2]   IUCN National Committee of The Netherlands, 1018 DD Amsterdam, The Netherlands;
      sander.vanandel@iucn.nl
[3]   Neukom Institute for Computational Science, Dartmouth College, Hanover, NH 03755, USA;
      jeffrey.t.kerby@dartmouth.edu
[4]   Conservation Science Department, the Jane Goodall Institute, 1595 Spring Hill Road, Suite 550, Vienna,
      VA 22182, USA; lpintea@janegoodall.org
[5]   Institute for Biodiversity and Ecosystem Dynamics, University of Amsterdam, Science Park 904,
      1098 XH Amsterdam, The Netherlands
[*]   Correspondence: N.Bonnin@2016.ljmu.ac.uk; Tel.: +44-7903-251803

Received: 6 March 2018; Accepted: 18 April 2018; Published: 23 April 2018

Abstract: As with other species of great apes, chimpanzee numbers have declined over the past decades. Proper conservation of the remaining chimpanzees requires accurate and frequent data on their distribution and density. In Tanzania, 75% of the chimpanzees live at low densities on land outside national parks and little is known about their distribution, density, behavior or ecology. Given the sheer scale of chimpanzee distribution across western Tanzania (>20,000 km^2), we need new methods that are time and cost efficient while providing precise and accurate data across broad spatial scales. Scientists have recently demonstrated the usefulness of drones for detecting wildlife, including apes. Whilst direct observation of chimpanzees is unlikely given their elusiveness, we investigated the potential of drones to detect chimpanzee nests in the Issa valley, western Tanzania. Between 2015 and 2016, we tested and compared the capabilities of two fixed-wing drones. We surveyed twenty-two plots (50 × 500 m) in gallery forests and miombo woodlands to compare nest observations from the ground with those from the air. We performed mixed-effects logistic regression models to evaluate the impact of image resolution, seasonality, vegetation type, nest height and color on nest detectability. An average of 10% of the nests spotted from the ground were detected from the air. From the factors tested, only image resolution significantly influenced nest detectability in drone-acquired images. We discuss the potential, but also the limitations, of this technology for determining chimpanzee distribution and density and to provide guidance for future investigations on the use of drones for ape population surveys. Combining traditional and novel technological methods of surveying allows more accurate collection of data on animal distribution and habitat connectivity that has important implications for ape conservation in an increasingly anthropogenically-disturbed landscape.

Keywords: UAV; great apes; conservation; survey; Tanzania; image resolution

1. Introduction

As with other great ape species, chimpanzee numbers have declined over the past decades and the species is currently threatened by extinction [1]. Several studies have documented the impact of habitat loss [2–4], poaching [5–7] and infectious disease [8,9] on wild populations. In Tanzania, 75% of wild chimpanzees are found within a 20,000 km^2 area of national parks [10–15]. Monitoring these chimpanzees is therefore crucial for their conservation in Tanzania. For conservation management, it is important to establish where and how many individuals remain and to understand the potential connectivity between populations. These data represent key information that is used towards creating baseline estimates for assessing the effectiveness of conservation efforts over time [16,17].

There are several established methods for studying and monitoring wild animal populations. Line transect surveys are widely used to estimate population density for a variety of mammal species, including great apes [18–21]. Data from direct observations of animals or indirect evidence such as dung [10], nests [22,23] and calls [24] can be converted into density and subsequently population estimates across larger landscapes [25]. Indirect evidence is especially important in great ape surveys given the elusive nature of the species and their extensive range and distribution [26].

Traditional land-based transects are time-consuming and expensive, and for these reasons geographically wide surveys are not repeated frequently [26]. Aerial surveys with light aircraft can be effective across broad areas for counting large mammals [27,28], but have limitations. While such surveys may provide an unbiased population size estimate for large mammals found in open areas (e.g., elephants, buffalos, zebras), they are unlikely to provide accurate estimates for smaller species (e.g., black-backed jackal, bushbuck, vervet monkey) [29] or those that live in habitats with greater canopy cover. Furthermore, aircraft surveys are logistically difficult to implement due to their very high cost and the risk they pose to operators (i.e., aircraft crashes) [30]. Due to their increasing availability, high resolution satellite images have also been used to detect animals or their signs [31]. Although promising, this method is also unlikely to provide accurate estimates for small species and is hampered by cost and atmospheric interference from clouds, especially problematic in the tropical regions where great apes are distributed [32]. Camera-traps and acoustic sensors are other promising remote technologies that enable broad spatiotemporal and precise information on animals that are elusive and otherwise difficult to study [33,34]. Nevertheless, these methods have high initial costs and still require intensive manual labor for deployment, memory card collection and substantial expertise in subsequent data analyses.

Recently, scientists have started to deploy drones—remotely operated aircraft with autonomous flight capabilities—for wildlife monitoring [35–37]. This application allows for rapid and frequent monitoring across moderate to broad spatial extents while providing high-resolution spatial data. Several studies have now reported successful animal detection using drone-derived aerial imagery, ranging from birds [36,38] to large terrestrial [39,40] and marine [41–44] mammals. Recent studies on using drones to detect indirect signs of animals have also reported promising results in detecting orangutan [45] as well as chimpanzee [46] nests.

Given the extent of the area in need of monitoring, exploring drone applications for chimpanzee population surveys in Tanzania may reduce cost and time investments. Visibility bias (i.e., failure to detect all animals within a sampled area) is a primary source of error in aerial surveys [27,29,47]. Prior to the widespread deployment of drones for a census, it is important to first evaluate bias in the method (i.e., calculate a correction factor) by comparing the resulting detections with traditional ground survey results. Numerous factors can impact the detectability of a direct or indirect sign of wildlife [25,48]. Thus, it is critical to determine what affects chimpanzee nest detectability in drone-acquired images. In the current study, we assessed several factors known to affect target detectability in aerial images: image resolution [39,49]; canopy cover and vegetation type [29,39,46,50]; and target size and color [29,42].

In summary, our objectives were to (1) evaluate drone performance for chimpanzee nest surveys by comparing ground and aerial surveys; and (2) assess the factors that influence detectability from drone data. Based on the results of previous studies, we hypothesized that using a higher resolution camera as well as flying at a lower altitude would increase the nest detection probability. We also expected a higher detection probability during the leaf-off season and in the more open miombo woodland vegetation than the closed riverine forest. Finally, we predicted that nests higher in the canopy and with a color that contrasts with their surroundings will be easier to detect.

2. Materials and Methods

2.1. Study Site

The study was conducted in May 2015 and September 2016 (beginning and end of dry seasons, respectively) in the Issa Valley, western Tanzania (Figures 1 and 2). The area is characterized by a landscape mosaic, dominated by miombo woodland (named for the dominant tree genera of *Brachystegia* and *Julbernardia*) interspersed with grasslands, swamps and gallery forest restricted to steep ravines. Open vegetation (e.g., miombo woodland, grassland and swamps) represents more than 90% of the 85 km² study area (Piel et al., unpublished data; Figure 1). The region is one of the driest, most open and seasonally extreme habitats in which chimpanzees live [51], with annual temperatures ranging from 11 °C to 35 °C and a dry season (<100 mm of rainfall) lasting from May to October.

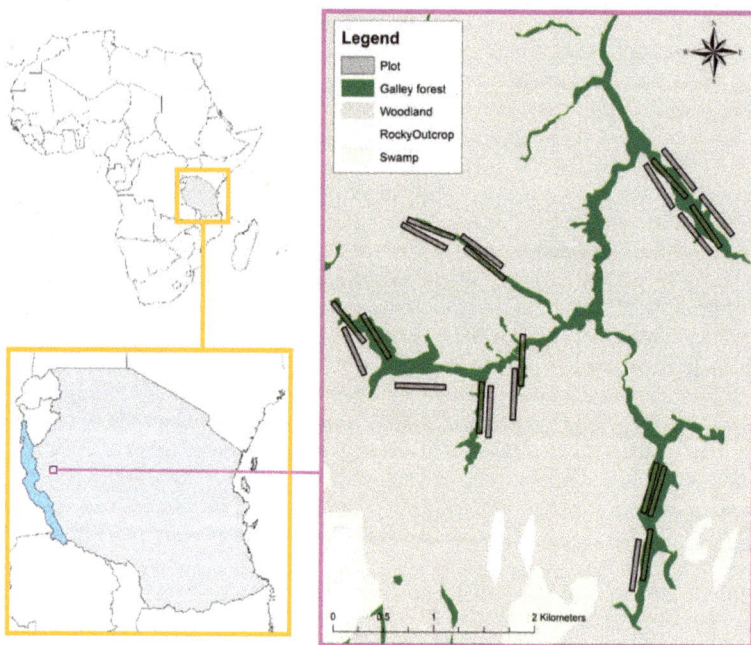

Figure 1. Location and map of the Issa Valley showing the distribution of all plots. Vegetation class layer produced by Caspian Johnson (unpublished).

Figure 2. Partial orthomosaics of the study site representative of the vegetation at the beginning (May 2015) and at the end (September 2016) of the dry season.

2.2. Ground Surveys

To collect chimpanzee nest data from the ground for comparison with drone observations, we created 22 plots, each 50 × 500 m, stratified equally across gallery forest and miombo woodland (Figure 1). Within each plot, two experienced observers walked slowly and recorded the GPS location of all observed chimpanzee nests. Only one inspection per plot was performed. During the 2015 survey, data were collected using the open data kit [52] on NEXUS 7 tablets with an average accuracy of 15 m. In 2016, we used the global navigation satellite system (GNSS) Mobile Mapper 20 (MM20, http://www.spectraprecision.com), allowing us to collect data with a <1 m accuracy. For each nest, we collected additional data, including nest height from ground (estimated to the nearest meter), vegetation type (open or closed) and the nest color (green or brown).

2.3. Aerial Surveys

For the aerial surveys, we used two drone models paired with two different cameras (Figure 3).

Figure 3. Types of drone/camera pairing deployed: (**a**) Pairing A; (**b**) Pairing B.

Pairing A: The ConservationDrones.org X5 (Skywalker X5 frame; hobbyking.com [similar to HBS FX61]) equipped with a GPS-enabled Canon S100 camera (resolution: 4000 × 3000 pixels; sensor size: 7.6 × 5.7 mm) operating a CHDK firmware modification.

Pairing B: The more stable HBS Skywalker 100 km Long Range Fix Wings drone (Skywalker 2013 body 1880 mm; hobbyking.com) fitted with a Sony RX100M2 (resolution: 5472 × 3648 pixels; sensor size: 13.2 × 8.8 mm). Both were equipped with an autopilot system based on the 'ArduPilot Mega' (APM), which includes a computer processor, GPS, data logger, pressure and temperature sensor, airspeed sensor, triple-axis gyro, and accelerometer. Cameras were triggered automatically based on a predefined flight plan to produce at least 60% front- and side-overlap among images. Missions were planned using the open-source software APM Mission Planner (http://planner.ardupilot.com/) on a standard Windows-based laptop. Once we completed the missions, we geotagged the images from the Sony camera using the same software. Geotagging was not necessary for the Canon images as the camera was GPS-equipped.

The drones performed two types of missions: straight line transects and grid missions (Figure 4).

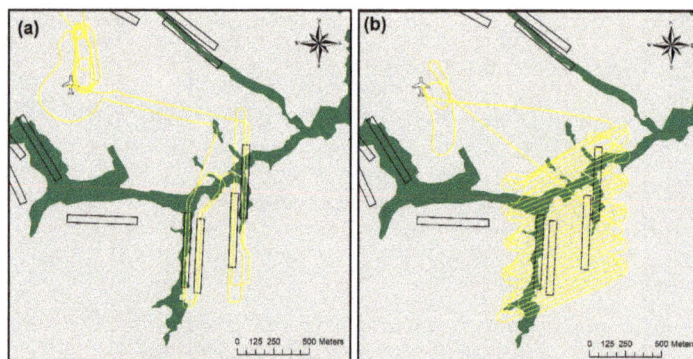

Figure 4. Types of mission flown: (**a**) Line transect; (**b**) Grid mission.

Line transects: Straight line missions covering the areas within the ground plots at an average altitude of 90 m above ground level (AGL). We investigated aerial images obtained during these missions for the presence of chimpanzee nests.

Grid missions: Grid pattern missions flown at an average altitude of 120 m above ground level with extensive overlap (>60%) between flight legs to allow for the creation of orthomosaics. We produced orthomosaics using the geotagged images in Pix4D mapper (https://pix4d.com, version 4.0.25). Although ground control points (GCPs) were set up in each area for both years, the GCPs from 2015 could not be localized in the aerial images. The resulting accuracy of the orthomosaics was that of the Canon S100 camera GPS (average accuracy of 5 m). Improved GCPs were set up in 2016 allowing a georeferencing accuracy within a meter. We used the orthomosaics for the subsequent spatial relocation of aerial observations made while interpreting the photos from the nest counting missions.

2.4. Nest Detection

One observer (NB) examined the 1227 images resulting from the transect missions falling within the plots. Images were imported into the WiMUAS software [53] and investigated for the presence of nests. The aerial observation location was subsequently exported to a georeferenced shapefile. Because the resulting file was accurate to within 50 m, each aerial observation was relocated using the orthomosaics. Due to the 15 m inaccuracy of the 2015 ground data, a buffer of 15 m was created around each nest and if an aerial observation was recorded within this 15 m radius that was considered an aerial nest detection.

2.5. Analyses

All statistical analyses were conducted in the R studio (version 1.0.136).

2.5.1. Performance of the Aerial Detection

We calculated recall and false alarm rates to estimate the performance of nest detection using drone imagery [54]. Recall is the percentage of successful detection (i.e., the proportion of nests observed from the ground detected during the aerial survey in relation to the total number of nests observed from the ground). The false alarm rate is the proportion of false detections (the number of aerial observations not aligning with nests found from the ground by the total number aerial observations). Because the data were not normally distributed, we used non-parametric statistics. A Wilcoxon-signed rank test was applied to compare the number of nests per plot found on the ground and on the aerial drone survey. We further ran a Spearman rank correlation to test for associations between the number of nests per plot across the two survey methods.

2.5.2. Factors Influencing Detectability

We used three generalized linear models (GLM) with a binomial error structure and logit-link function to evaluate which factors (drone/camera pairing, season, vegetation type, nest age, nest height and flight altitude above ground level (AGL)) influenced the recall rate and the false alarm rate. The models were fitted using the GLM function from the lme4 package [55]. We fitted all terms of interest and tested significance via likelihood ratio tests to determine which factors resulted in a significant reduction in explanatory power when removed [56].

Factors influencing the recall rate: For the first model, the recall rate was fitted following the method from Lopez-Bao [57]. The number of nest detection successes vs. number of failures by plot (modelled as 1 = success and 0 = failure) was fitted as the dependent variable. Drone/camera pairing (Pairing A or Pairing B), season (May 2015 or September 2016) and vegetation type (open or closed) were each fitted as two-level fixed effects. As it was not possible to test the influence of all variables in this model (e.g., nest color and nest height required a perfect individual nest match between the ground and aerial survey), we fitted a second model. This second model included only the data from the 2016 survey, for which aerial observations could be more accurately matched to individual nests found on the ground. We fitted the nest detection event (not detected = 0, detected = 1) as the dependent variable. Vegetation type (open vs. closed) and nest color (green or brown) were each fitted as the two-level fixed effect and flight altitude AGL and nest height were fitted as covariates. We determined flight altitude AGL by subtracting the elevation (extracted from a Shuttle Radar Topographic Mission (SRTM) layer—30 m resolution; http://earthexplorer.usgs.gov) from the flight altitude above mean sea level (extracted from the geotagged images) at each recorded nest location.

Factors influencing the false alarm rate: In the last model, the false detection event (true detection = 0, false detection = 1) was fitted as dependent variable. Drone/camera pairing (Pairing A or Pairing B), season (May 2015 or September 2016) and vegetation type (open or closed) were each fitted as two-level fixed effects and flight altitude AGL was fitted as a covariate.

3. Results

3.1. Performance of the Aerial Detection

Considering both survey seasons (May 2015 and September 2016) and the results from both drone/camera pairings (pairing A and pairing B), we documented 667 chimpanzee nests from the ground (Supplementary Figure S1) and 112 from aerial observations (Figure 5; Supplementary Figure S2). Of these aerial observations, 64 fell within the 15 m radius of a nest that had been spotted from the ground and were considered to be nests, representing a 9.6% recall rate and 42.8% false alarm rate. Although the image analysis resulted in significantly fewer nest records per plot compared to what the ground teams documented (Wilcoxon- signed rank test: $v = 981$; $p < 0.001$; $n = 47$), the number of nests detected from aerial survey imagery showed a significantly positive correlation with those recorded on the ground per plot (Spearman's $\rho = 0.53$; $p < 0.001$, $n = 47$).

Figure 5. Examples of images of chimpanzee nests: captured during drone surveys (**a,b**) and observed from the ground (**c,d**).

3.2. Factors Influencing Detectability

3.2.1. Factors Influencing the Recall Rate

Our first model included drone/camera pairing and season and vegetation type. From these variables, only drone/camera pairing significantly influenced the recall rate (likelihood ratio test: $X^2 = -10.96, p < 0.001$), with the highest probability of nest detection with Pairing B (12.81% probability) (Figure 6). There was no significant difference in the recall rate between open and closed vegetation types (likelihood ratio test: $X^2 = 93.1$, df $= 41$, $p = 0.747$) or between the beginning and end of the dry season (likelihood ratio test: $X^2 = 93$, df $= 43$, $p = 0.551$) (Table 1).

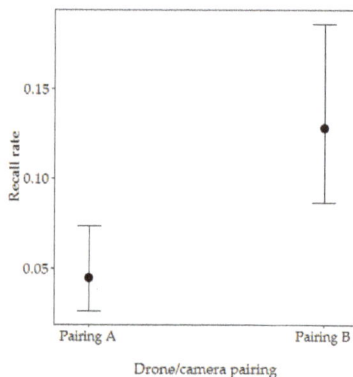

Figure 6. Effect of drone/camera pairing on the recall rate. Error bars represent 95% confidence intervals for the predicted probabilities.

Table 1. Outcomes of GLM to investigate the effect of drone/camera pairing, season and vegetation on the recall rate.

Predictors	LRT		Parameter Estimate			
	X^2	*p* Value	Estimate	Std. E.	*z* Value	Pr (>|z|)
(Intercept)			−2.96	0.59	−5.01	5.66×10^{-7}
Drone/camera pairing (Pairing A) Pairing B	10.96	0.004 **	1.43	0.57	−2.49	0.013 *
Vegetation (closed) Open	0.89	0.828	0.3	0.84	0.37	0.722
Season (May 2015) Sep-16	0.40	0.818	−0.35	0.78	−0.45	0.651
Drone/camera pairing: Vegetation Pairing A: Open vegetation	0.55	0.457	0.57	0.76	0.74	0.458
Vegetation: Season Open vegetation: September 2016	7.29	0.993	0.01	1	0.01	0.993

The *p* value for each term is based on the chi-squared test (likelihood ratio test (LRT)) for change in the deviance when comparing models with or without that term. Parameter estimates are reported for all terms in the full model. * = *p* < 0.05; ** = *p* < 0.01.

Our second model (for 2016 data only) included flight altitude, nest height and vegetation type. We decided to remove nest color from our second model as of the 337 nests recorded by the ground survey team in 2016, only one was green. The recall rate differed significantly across flight altitude AGL (likelihood ratio test: $X^2 = 4.35$, $p < 0.05$), with nests more likely to be detected when flying at a lower altitude (19.58% probability) (Figure 7). We found a trend towards higher detectability in closed rather than open vegetation (likelihood ratio test: $X^2 = 2.79$, $p < 0.1$) (Table 2). There was no significant difference in nest detection depending on nest height within the tree (likelihood ratio test: $X^2 = 0.07$, $p = 0.789$).

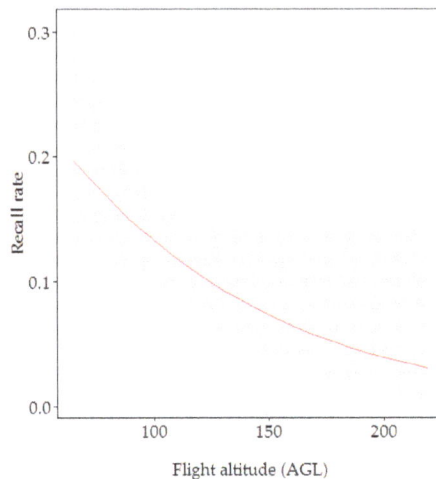

Figure 7. Effect of the flight altitude (AGL) on the recall rate. Grey ribbon represents 95% confidence intervals for predicted probabilities.

Table 2. Outcomes of GLM to investigate the effect of altitude, vegetation type and nest height on the recall rate.

Predictors	LRT		Parameter Estimate			
	X^2	*p* Value	Estimate	Std. E.	*z* Value	Pr (>|z|)
(Intercept)			−1.53	0.28	−5.45	4.98×10^{-8}
Flight altitude AGL	4.35	0.037 *	−0.47	0.25	−1.90	0.057
Vegetation (closed) Open	2.79	0.094	−0.68	0.40	−1.70	0.089
Nest height	0.07	0.789	0.04	0.17	0.27	0.789

The *p* value for each term is based on the chi-squared test (likelihood ratio test (LRT)) for change in deviance when comparing models with or without that term. Parameter estimates are reported for all terms in the full model. * = *p* < 0.05.

3.2.2. Factors Influencing the False Alarm Rate

For this model, we investigated the influence of drone/camera pairing, season, vegetation type and flight altitude AGL on the false alarm rate. Drone/camera pairing, vegetation type and flight altitude AGL significantly influenced the false alarm rate (Table 3). Aerial observations from Pairing A were more likely to be false positives (0.83% probability). The overall false alarm rate was higher in closed vegetation than in open vegetation but significantly differed between seasons (likelihood ratio test: X^2 = 4.01, *p* < 0.05). Aerial observations made at the beginning of the dry season (May 2015) were more likely to be false positives when recorded in open vegetation (0.94% probability opposed to 0.19% probability for closed vegetation). The false alarm rate significantly increased at lower altitude (likelihood ratio test: X^2 = 9.55, *p* < 0.05) (Figure 8).

Table 3. Outcomes of GLM investigating the effect of the drone/camera pairing, season, vegetation type and flight altitude AGL on the false alarm rate.

Predictors	LRT		Parameter Estimate			
	X^2	*p* Value	Estimate	Std. E.	*z* Value	Pr (>|z|)
(Intercept)			−3.03	1.19	−2.54	0.011 *
Drone/camera pairing (Pairing A) Pairing B	14.14	1.17×10^{-4} ***	3.69	1.08	3.40	6.73×10^{-4} ***
Vegetation (closed) Open	23.23	1.44×10^{-6} ***	5.72	1.99	2.87	0.004 **
Season (May 2015) Sep-16	0.04	0.834	2.86	1.16	2.47	0.013 *
Flight altitude AGL	9.55	0.002 **	2.01	0.90	2.24	0.025 *
Drone/camera pairing: Vegetation Pairing A: Open vegetation	0.05	0.824	−3.72	1.56	−2.38	0.017 *
Season: Vegetation Sept 2016: Open vegetation	4.01	0.045 *	−7.27	1.83	−3.98	6.83×10^{-5} ***
Vegetation: Flight altitude AGL Open vegetation: Flight altitude AGL	0.37	0.542	−5.98	1.63	−3.67	2.40×10^{-4} ***

The *p* value for each term is based on the chi-squared test (likelihood ratio test (LRT)) for change in deviance when comparing models with or without that term. Parameter estimates are reported for all terms in the full model. * = *p* < 0.05; ** = *p* < 0.01; *** = *p* < 0.001.

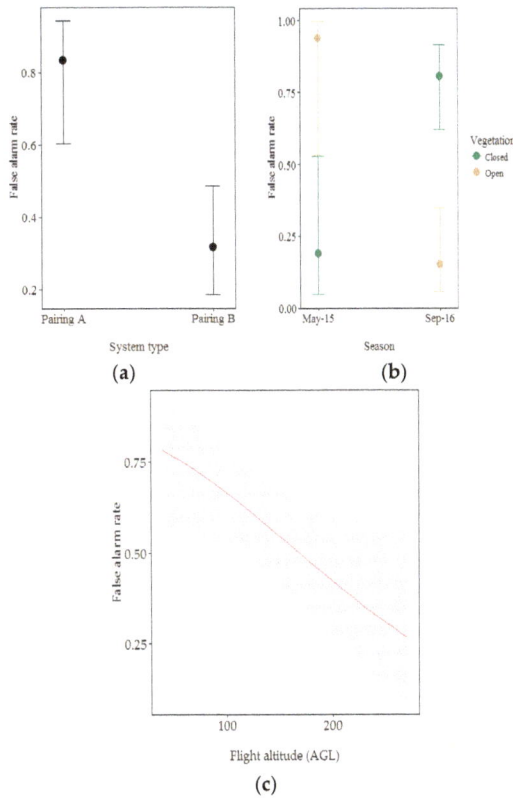

Figure 8. Effect of (**a**) drone/camera pairing; (**b**) vegetation type within season and (**c**) flight altitude above ground level (AGL) on the false alarm rate. Error bars and grey ribbon represent 95% confidence intervals for predicted probabilities.

4. Discussion

We investigated the feasibility of using drones to detect chimpanzee nests in the Issa Valley, western Tanzania, and evaluated the influence of image resolution, seasonality, vegetation type, nest height and color on nest detectability. An average of 10% of the nests observed from the ground were detected from the air, with improved nest detection in imagery with higher spatial resolution. Our overall detection rate was lower than that previously reported for chimpanzee nests in Gabon (39.9%) [46] and orangutan nests in Indonesia (17.4%) [45]. This discrepancy is likely due to methodological differences and our systematic approach. In their study, van Andel et al. [46] used two approaches that biased the probability of detection. In the first, they collected nest data first via ground surveys and then used the location of the recorded nests to confirm their presence in drone images. In the second, nests were first detected on drone images and then confirmed on the ground using the location of the aerial observations. These methods effectively demonstrated that it was indeed possible to detect chimpanzee nests from drones, although these specific approaches resulted in an increased probability of detecting a nest in the drone images for the first approach and on the ground for the second approach. Wich et al. [45] used a buffer of 25 m around nests recorded on the ground to select which nest detected from the air would be included in the analyses, comparing the relative density of nests from the aerial and ground-based surveys. The smaller 15 m buffer used in our study could be associated with our smaller detection rate, i.e., we were more conservative regarding what constituted a match. Moreover, aerial nest surveys may be more efficient for orangutan nests as they tend to build nests higher in the tree canopy and visual contrasts of nest materials and canopy color are seemingly more apparent in these habitats [58,59].

Of the factors hypothesized to influence the probability of chimpanzee nest detection in drone-derived aerial imagery, only image resolution was identified as having a significant influence on the recall rate, with a higher probability of nest detection associated with the higher-resolution camera at a lower flight altitude AGL. This finding is consistent with that of Mulero-Pázmány et al. [39], who also found that the targets (i.e., rhinoceros, people acting as poachers) were better detected with a lower-flying drone. Our results are also consistent with those of Dulava et al. [49], who reported a significant negative relation between ground sampling distance (GSD) and correct waterbird identification with a minimum of 5 mm GSD. In our study, we favored flight altitude AGL above GSD as a measure of resolution because of identical camera parameters, however, the two are conceptually interchangeable. We obtained the highest probability of nest detection at the lowest possible flight altitude (AGL: 65 m), corresponding to 1.4 cm GSD. Flying at lower altitude would have threatened drone safety. These findings reflect the inherent trade-offs between monitoring at a high spatial resolution (grain) versus across broad spatial extents, such as ground sampling distance (GSD) and ground sampling area (GSA), scale inversely with one another. This highlights the importance of the a priori identification of the minimum GSD required to detect ground targets from the air during the survey design period, particularly if planning for extensive area surveys where the balance between GSD and GSA should be optimized.

Contrary to expectations, we did not find a significant influence of nest height on aerial nest detection. Nests constructed higher in trees are expected to be more visible from the air, however, the visibility also depends on the height of the tree (i.e., a nest at 15 m will be more visible in a tree of 15 m height than in a tree of 20 m). The inclusion of tree height into models will be important in subsequent analyses.

Another surprising result of our study was the lack of influence of canopy cover and vegetation type, with no significant differences between the probability of nest detection in the leaf-off season and the "greener season" as well as between the more open miombo woodland vegetation and the closed, riverine forest. Even more surprising, the probability of nest detection tended to be higher in closed rather than in open vegetation. This finding contradicts numerous other studies that demonstrated a significant improvement of target detection from drone imagery in more open habitats (e.g., [29,39,46,50,60]). A possible explanation for this might be the difficulty of detecting brown nests against a similarly colored background, in this case the less continuous and more earth-toned colors of the Miombo woodland and the grassland mosaic. Light body color has been demonstrated to negatively influence animal detection during aerial surveys in a conservation area of northern Tanzania (e.g., dark Ostrich (*Struthio camelus*) better detected than light Grant's gazelle (*Nanger granti*)) [29]. The results from Chabot and Bird [61] further support the importance of contrast in target detection. In their investigation into the use drones to survey flocks of geese they reported a poor detection of low-contrast Canada Geese (*Branta canadensis*) but good aerial survey performance for the high-contrast Snow Geese (*Chen caerulescens*) resulting in more efficient aerial count compared to ground count (60% higher). We were unable to test the role of contrast in our study due to an insufficient sample of recent (green) nests.

Findings from the analysis of the factors influencing false alarm rates support this hypothesis. Different vegetation types significantly affected the false alarm rate depending on the season. The false alarm rate was higher in miombo woodland at the beginning of the dry season. The canopy cover in miombo woodland is much higher during this period than at the end of the dry season. At the beginning of the dry season, the miombo woodland reflects a mosaic of green leaves and a brown understory, leading to potential misinterpretation of aerial data. At the end of the dry season, however, reflection is mostly from the brown understory, making nest detection more difficult but more accurate.

As only Paring A was flown in both seasons, we acknowledge that technological factors may play a role in these seasonal effects, however we strongly believe future studies will benefit from considering and further exploring the effects of seasonal canopy differences on nest detection.

The limitations on the use of drones to survey chimpanzees are threefold. Firstly, only a small proportion of chimpanzee nests are detectable from the air. Most chimpanzee nests are built in the middle of the tree crown [62] making them undetectable from above the tree canopy [46]. Chimpanzees also exhibit ground night nesting [63], which would also be difficult to detect from aerial surveys. Secondly, the high proportion of false alarm rate highlighted in this study is problematic. The false alarm rate is an important parameter that must be taken into consideration when assessing new wildlife survey methods, as it may lead to an overestimation of the population density [29]. However, the false alarm rate has not been described in previous studies investigating the use of drones to detect great ape nests. In this study, we reported a 42.8% false alarm rate. These aerial observations, for which the location did not align with any of the nests spotted from the ground, can be explained in two ways: (1) These could be nests visible from the air, but not the ground, as would be the case of nests high in the canopy that might be obscured from ground teams by the mid-canopy. Van Schaik et al. [64] noted that nests can go undetected during ground surveys, resulting in an underestimation of ape densities; (2) alternatively, false positives could represent dead leaves or canopy gaps revealing the brown understory that was mistaken for nests. This uncertainty represents an important problem in the deployment of drones to assess chimpanzee presence/density, especially in a new area where little information is available. We argue here that whilst aerial imagery offers an improvement in spatial coverage and data collection time and frequency, this approach still requires complimentary validation from ground surveys. Finally, the time associated with analyzing thousands of images to identify nests represents an additional key limitation to using drones in this context.

The limitations we discussed above are significant but not prohibitive, and the findings from our study provide guidance for future investigation on the use of drones for ape population surveys. Firstly, it is important to generate high spatial resolution images and lower GSD, providing greater details and significantly increasing the probability of nest detection. For our survey, we decided to use fixed-wing drone models allowing longer flights that can cover larger areas. Because of the mountainous terrain, flying at lower altitude was not possible. Most chimpanzees do not live across mountainous terrain, therefore this problem would not affect large parts of their range. Multirotor drones have smaller flight time capacities but can fly at lower altitudes [65]. This technology is improving rapidly (e.g., drone design optimization allowing longer flight time [66,67]), which could make multirotors a viable option in the future. Meanwhile, camera resolution is improving, which will allow future studies to obtain higher resolution images from fixed wing surveys. Reliable detection also requires a high contrast background. During both our survey seasons, the brown understory made nest identification difficult. We therefore recommend conducting future surveys during seasons with green vegetation on the ground to contrast otherwise brown nests. We acknowledge that this context might reduce the probability of detecting fresh green nests, however, given their low abundance, their non-detectability is less likely to impact chimpanzee density estimations. Multispectral sensors may help address this problem. Widely used for landcover classification and vegetation monitoring [68–73] this technology uses green, red, red-edge and near infrared wavebands to capture detail not available to standard RGB cameras. Green vegetation materials are characterized by high reflectance in the near infra-red (NIR) domain (outside of the spectral range of human vision); a multispectral camera can provide useful contrast to discriminate between live and dead vegetation. Furthermore, it would be interesting to assess the potential of oblique aerial images. This perspective may offer better glimpses through foliage and more intuitively interpretable representations of the targets. Another step would be to assess the potential of 3D mapping of the canopy surface for nest detection. 3D models can now be created using point clouds from drone imagery [74] providing better perspectives for visual interpretation of the data. Another complimentary approach would be to use light detection and ranging (LiDAR) technology. Recently developed at sizes suitable for drone payloads [75], this remote sensing technique offers new insights beyond simple top of canopy structure that may help nest detectability algorithms. For example, these technologies could be used to better establish the habitat characteristics of trees holding nests. These data could be used in computer vision algorithms [76–79] to refine automatic nest detection, possibly reducing the false alarm rate. A recent study on using a drone to detect eagle nests reported 75% nest detection using a semi-automated method [80]. Similar to the difficulties encountered with chimpanzee nest detection, eagle nests are found in highly heterogeneous environments with many features that resemble nests, at small scale (~1–2 m), and with variable nest size, shape and context. This result is promising for broader nest detection applications, including those of great apes.

Given the shy and elusive nature of great apes, direct surveys are rarely feasible. Researchers thus must rely on indirect signs to estimate population density. However, to convert nest counts into ape density, the nest decay rate and nest production rate are required. These factors are highly dependent on apes species and environmental characteristics, and therefore require extensive study [26]. Recent studies have now shown the potential of thermal cameras mounted on drones for animal detection [39,76,81]. However, this approach would require extensive spatial coverage and further research is required to assess whether apes could be detected using a thermal camera mounted on a drone.

5. Conclusions

The design and execution of great ape surveys are crucial to allocating conservation efforts to where they are most needed, but face many logistical challenges, particularly when they must be implemented across broad areas. Drone surveys could be a revolutionary method, allowing rapid and frequent monitoring in remote and poorly-understood areas, with data accessible immediately and containing a rich variety of information about habitat and other conservation revelation conditions. The limitations we discussed above are meaningful but not prohibitive, and the rapid pace of technological improvement suggests many promising solutions in a near future. Assessing the potential of drones to detect chimpanzee nests has major implications, not only for chimpanzee monitoring across Tanzania, but also for all great apes monitoring. This technology could be applied to survey extensive areas filling problematic gaps in our current understanding of ape distribution and abundance [82], providing key information for conservationists.

Supplementary Materials: The following are available online at http://www.mdpi.com/2504-446X/2/2/17/s1, Figure S1: Locations of nests observed from the ground, Figure S2: Aerial observations (true positives and false positives) recorded from drone surveys.

Acknowledgments: We thank the Tanzanian Wildlife Research Institute (TAWIRI) and Commission for Science and Technology (COSTECH) for permission to carry out research in Tanzania. This work was supported by Tuungane/The Nature Conservancy, the Jane Goodall Institute (JGI) with additional support through the USAID and NASA's Ecological Forecasting for Conservation and Natural Resource Management program and National Geographic. Special thanks to DigitalGlobe and Esri for inkind contribution and support with satellite imagery and GIS software. We are also extremely grateful to Mashaka Alimas, Jonas Bukende, Adrienne Chitayat, Patrick Hassan, Mlela Juma, Shedrack Lucas, Msigwa Rashid, Godfrey Stephano, Eden Wondra and Elizabeth Yarwood for field assistance. We thank Keeyen Pang from Hornbillsurveys for building the drones and his support. Many thanks to James Waterman for his statistical advice. Finally, thank you to Anne-Sophie Crunchant, Camille Guiliano and Ineke Knot for their valuable feedback on the manuscript.

Author Contributions: S.A.W., A.K.P., A.C.V.A. and J.T.K. conceived and designed the study; N.B., J.T.K. and A.C.V.A. collected the data; N.B. analyzed the data; N.B. wrote the manuscript with editorial contributions from S.A.W., A.K.P., J.T.K. and A.C.V.A.

Conflicts of Interest: The authors declare no conflict of interest.

References

1. International Union for Conservation of Nature (IUCN). The IUCN Red List of Threatened Species. Version 2017-2. 2017. Available online: http://www.iucnredlist.org (accessed on 21 September 2017).

2. Campbell, G.; Kuehl, H.; N'Goran Kouamé, P.; Boesch, C. Alarming decline of West African chimpanzees in Côte d'Ivoire. *Curr. Biol.* **2008**, *18*, 903–904. [CrossRef] [PubMed]

3. Junker, J.; Blake, S.; Boesch, C.; Campbell, G.; du Toit, L.; Duvall, C.; Ekobo, A.; Etoga, G.; Galat-Luong, A.; Gamys, J.; et al. Recent decline in suitable environmental conditions for African great apes. *Divers. Distrib.* **2012**, *18*, 1077–1091. [CrossRef]

4. Wich, S.A.; Garcia-Ulloa, J.; Kühl, H.S.; Humle, T.; Lee, J.S.H.; Koh, L.P. Will oil palm's homecoming spell doom for Africa's great apes? *Curr. Biol.* **2014**, *24*, 1659–1663. [CrossRef] [PubMed]

5. Bowen-Jones, E.; Pendry, S. The threat to primates and other mammals from the bushmeat trade in Africa, and how this threat could be diminished. *Oryx* **1999**, *33*, 233–246. [CrossRef]

6. McLennan, M.R.; Hyeroba, D.; Asiimwe, C.; Reynolds, V.; Wallis, J. Chimpanzees in mantraps: lethal crop protection and conservation in Uganda. *Oryx* **2012**, *46*, 598–603. [CrossRef]

7. Piel, A.K.; Lenoel, A.; Johnson, C.; Stewart, F.A. Deterring poaching in western Tanzania: The presence of wildlife researchers. *Glob. Ecol. Conserv.* **2015**, *3*, 188–199. [CrossRef]

8. Walsh, P.D.; Abernethy, K.A.; Bermejo, M.; Beyers, R.; De Wachter, P.; Akou, M.E.; Huijbregts, B.; Mambounga, D.I.; Toham, A.K.; Kilbourn, A.M.; et al. Catastrophic ape decline in western equatorial Africa. *Nature* **2003**, *422*, 611–614. [CrossRef] [PubMed]

9. Rudicell, R.S.; Holland Jones, J.; Wroblewski, E.E.; Learn, G.H.; Li, Y.; Robertson, J.D.; Greengrass, E.; Grossmann, F.; Kamenya, S.; Pintea, L.; et al. Impact of simian immunodeficiency virus infection on chimpanzee population dynamics. *PLoS Pathog.* **2010**, *6*, e1001116. [CrossRef] [PubMed]

10. Moore, D.L.; Vigilant, L. A population estimate of chimpanzees (*Pan troglodytes schweinfurthii*) in the Ugalla region using standard and spatially explicit genetic capture—recapture methods. *Am. J. Primatol.* **2013**. [CrossRef] [PubMed]

11. Plumptre, A.J.; Rose, R.; Nangendo, G.; Williamson, E.A.; Didier, K.; Hart, J.; Mulindahabi, F.; Hicks, C.; Griffin, B.; Ogawa, H.; et al. *Eastern Chimpanzee (Pan troglodytes schweinfurthii) Status Survey and Conservation Action Plan 2010–2020*; International Union for Conservation of Nature (IUCN): Gland, Switzerland, 2010.

12. Piel, A.K.; Stewart, F.A. *Census and Conservation Status of Chimpanzees (Pan troglodytes schweinfurthii) Across the Greater Mahale Ecosystem*; The Nature Conservancy: Arlington, VA, USA, 2014; 74p.

13. Kano, T.; Ogawa, H.; Asato, R.; Kanamori, M. Distribution and density of wild chimpanzees on the northwestern bank of the Malagarasi River, Tanzania. *Primate Res.* **1999**, *15*, 153–162. [CrossRef]

14. Ogawa, H.; Yoshikawa, M.; Mbalamwezi, M. A Chimpanzee bed found at Tubila, 20 km from Lilanshimba habitat. *Pan Africa News* **2011**, *18*, 5–6. [CrossRef]

15. Zamma, K.; Inoue, E. On the chimpanzees of Kakungu, Karobwa and Ntakata. *Pan Africa News* **2004**, *11*, 8–10. [CrossRef]

16. Plumptre, A.J.; Cox, D. Counting primates for conservation: Primate surveys in Uganda. *Primates* **2006**, *47*, 65–73. [CrossRef] [PubMed]

17. Nichols, J.D.; Williams, B.K. Monitoring for conservation. *Trends Ecol. Evol.* **2006**, *21*, 668–673. [CrossRef] [PubMed]

18. Silveira, L.; Jácomo, A.T.A; Diniz-Filho, J.A.F. Camera trap, line transect census and track surveys: A comparative evaluation. *Biol. Conserv.* **2003**, *114*, 351–355. [CrossRef]

19. Piel, A.K.; Cohen, N.; Kamenya, S.; Ndimuligo, S.A.; Pintea, L.; Stewart, F.A. Population status of chimpanzees in the Masito-Ugalla Ecosystem, Tanzania. *Am. J. Primatol.* **2015**, *77*, 1027–1035. [CrossRef] [PubMed]

20. Wich, S.A.; Singleton, I.; Nowak, M.G.; Utami Atmoko, S.S.; Nisam, G.; Arif, S.M.; Putra, R.H.; Ardi, R.; Fredriksson, G.; Usher, G.; et al. Land-cover changes predict steep declines for the Sumatran orangutan (Pongo abelii). *Sci. Adv.* **2016**, *2*, e1500789. [CrossRef] [PubMed]

21. Stokes, E.J.; Strindberg, S.; Bakabana, P.C.; Elkan, P.W.; Iyenguet, F.C.; Madzoké, B.; Malanda, G.A.F.; Mowawa, B.S.; Moukoumbou, C.; Ouakabadio, F.K.; et al. Monitoring great ape and elephant abundance at large spatial scales: Measuring effectiveness of a conservation landscape. *PLoS ONE* **2010**, *5*, e10294. [CrossRef] [PubMed]

22. Kouakou, C.Y.; Boesch, C.; Kuehl, H. Estimating chimpanzee population size with nest counts: Validating methods in Ta? National Park. *Am. J. Primatol.* **2009**, *71*, 447–457. [CrossRef] [PubMed]
23. Spehar, S.N.; Mathewson, P.D.; Nuzuar; Wich, S.A.; Marshall, A.J.; Kühl, H.; Nardiyono; Meijaard, E. Estimating orangutan densities using the standing crop and marked nest count methods: Lessons learned for conservation. *Biotropica* **2010**, *42*, 748–757. [CrossRef]
24. Kidney, D.; Rawson, B.M.; Borchers, D.L.; Stevenson, B.C.; Marques, T.A.; Thomas, L. An efficient acoustic density estimation method with human detectors applied to gibbons in Cambodia. *PLoS ONE* **2016**, *11*, e155066. [CrossRef] [PubMed]
25. Buckland, S.T.; Anderson, D.R.; Burnham, K.P.; Laake, J.L.; Borchers, D.L.; Thomas, L. *Introduction to Distance Sampling: Estimating Abundance of Biological Populations*; Oxford University Press: Oxford, UK, 2001.
26. Kühl, H.; Maisels, F.; Ancrenaz, M.; Williamson, E.A. *Best Practice Guidelines for Surveys and Monitoring of Great Ape Populations*; International Union for Conservation of Nature (IUCN): Gland, Switzerland, 2009.
27. Jachmann, H. Comparison of aerial counts with ground counts for large African herbivores. *J. Appl. Ecol.* **2002**, *39*, 841–852. [CrossRef]
28. Kirkman, S.P.; Yemane, D.; Oosthuizen, W.H.; Meÿer, M.A.; Kotze, P.G.H.; Skrypzeck, H.; Vaz Velho, F.; Underhill, L.G. Spatio-temporal shifts of the dynamic Cape fur seal population in Southern Africa, based on aerial censuses (1972–2009). *Mar. Mammal Sci.* **2013**, *29*, 497–524. [CrossRef]
29. Greene, K.; Bell, D.; Kioko, J.; Kiffner, C. Performance of ground-based and aerial survey methods for monitoring wildlife assemblages in a conservation area of northern Tanzania. *Eur. J. Wildl. Res.* **2017**, *63*, 77. [CrossRef]
30. Sasse, D.B. Job-related mortality of wildlife workers in the United States, 1937–2000. *Wildl. Soc. Bull.* **2003**, *31*, 1000–1003.
31. Yang, Z.; Wang, T.; Skidmore, A.K.; De Leeuw, J.; Said, M.Y.; Freer, J. Spotting East African mammals in open savannah from space. *PLoS ONE* **2014**, *9*, e115989. [CrossRef] [PubMed]
32. Hansen, M.C.; Roy, D.P.; Lindquist, E.; Adusei, B.; Justice, C.O.; Altstatt, A. A method for integrating MODIS and Landsat data for systematic monitoring of forest cover and change in the Congo Basin. *Remote Sens. Environ.* **2008**, *112*, 2495–2513. [CrossRef]
33. Rowcliffe, J.M.; Carbone, C. Surveys using camera traps: Are we looking to a brighter future? *Anim. Conserv.* **2008**, *11*, 185–186. [CrossRef]
34. Blumstein, D.T.; Mennill, D.J.; Clemins, P.; Girod, L.; Yao, K.; Patricelli, G.; Deppe, J.L.; Krakauer, A.H.; Clark, C.; Cortopassi, K.A.; et al. Acoustic monitoring in terrestrial environments using microphone arrays: Applications, technological considerations and prospectus. *J. Appl. Ecol.* **2011**, *48*, 758–767. [CrossRef]
35. Koh, L.P.; Wich, S.A. Dawn of drone ecology: low-cost autonomous aerial vehicles for conservation. *Trop. Conserv. Sci.* **2012**, *5*, 121–132. [CrossRef]
36. Chabot, D.; Bird, D.M. Wildlife research and management methods in the 21st century: Where do unmanned aircraft fit in? *J. Unmanned Veh. Syst.* **2015**, *3*, 137–155. [CrossRef]
37. Wich, S.A. Drones and conservation. In *Drones and Aerial Observation: New Technologies for Property Rights, Human Rights, and Global Development. A Primer*; Kakaes, K., Ed.; New America: Washington, DC, USA, 2015; pp. 63–71.
38. Chabot, D.; Carignan, V.; Bird, D.M. Measuring habitat quality for least bitterns in a created wetland with use of a small unmanned aircraft. *Wetlands* **2014**, *34*, 527–533. [CrossRef]
39. Mulero-Pázmány, M.; Stolper, R.; Van Essen, L.D.; Negro, J.J.; Sassen, T. Remotely piloted aircraft systems as a rhinoceros anti-poaching tool in Africa. *PLoS ONE* **2014**, *9*, e83873. [CrossRef] [PubMed]
40. Vermeulen, C.; Lejeune, P.; Lisein, J.; Sawadogo, P.; Bouché, P. Unmanned aerial survey of elephants. *PLoS ONE* **2013**, *8*, e54700. [CrossRef] [PubMed]
41. Hodgson, A.J.; Kelly, N.; Peel, D. Unmanned aerial vehicles (UAVs) for surveying Marine Fauna: A dugong case study. *PLoS ONE* **2013**, *8*, e79556. [CrossRef] [PubMed]
42. Koski, W.R.; Allen, T.; Ireland, D.; Buck, G.; Smith, P.R.; Macrender, A.M.; Halick, M.A.; Rushing, C.; Sliwa, D.J.; McDonald, T.L. Evaluation of an unmanned airborne system for monitoring marine mammals. *Aquat. Mamm.* **2009**, *35*, 347–357. [CrossRef]
43. Koski, W.R.; Gamage, G.; Davis, A.R.; Mathews, T.; LeBlanc, B.; Ferguson, S.H. Evaluation of UAS for photographic re-identification of bowhead whales, Balaena mysticetus. *J. Unmanned Veh. Syst.* **2015**, *3*, 22–29. [CrossRef]

44. Hodgson, A.; Peel, D.; Kelly, N. Unmanned aerial vehicles for surveying marine fauna: Assessing detection probability. *Ecol. Appl.* **2017**, *27*, 1253–1267. [CrossRef] [PubMed]

45. Wich, S.A.; Dellatore, D.; Houghton, M.; Ardi, R.; Koh, L.P. A preliminary assessment of using conservation drones for Sumatran orang-utan (*Pongo abelii*) distribution and density. *J. Unmanned Veh. Syst.* **2015**, *4*, 45–52. [CrossRef]

46. Van Andel, A.C.; Wich, S.A.; Boesch, C.; Koh, L.P.; Robbins, M.M.; Kelly, J.; Kuehl, H.S. Locating chimpanzee nests and identifying fruiting trees with an unmanned aerial vehicle. *Am. J. Primatol.* **2015**, *77*, 1122–1134. [CrossRef] [PubMed]

47. Pollock, K.H.; Kendall, W.L. Visibility bias in aerial surveys: A review of estimation procedures. *J. Wildl. Manag.* **1987**, *51*, 502–510. [CrossRef]

48. Buckland, S.; Anderson, D.R.; Burnham, K.; Laake, J.; Borchers, D.; Thomas, L. *Advanced Distance Sampling: Estimating Abundance of Biological Populations*; Oxford University Press: Oxford, UK, 2004.

49. Dulava, S.; Bean, W.T.; Richmond, O.M.W. Environmental reviews and case studies: Applications of unmanned aircraft systems (UAS) for waterbird surveys. *Environ. Pract.* **2015**, *17*, 201–210. [CrossRef]

50. Patterson, C.; Koski, W.; Pace, P.; Mcluckie, B.; Bird, D.M. Evaluation of an unmanned aircraft system for detecting surrogate caribou targets in Labrador. *J. Unmanned Veh. Syst.* **2016**, *4*, 53–69. [CrossRef]

51. Moore, J. Savanna chimpanzees. In *Topics in Primatology, Vol.1 Human Origins*; Nishida, T., McGrew, P., Marler, P., Pickford, M., de Waal, F., Eds.; University of Tokyo Press: Tokyo, Japan, 1992; pp. 99–118.

52. Anokwa, Y.; Hartung, C.; Brunette, W.; Borriello, G.; Lerer, A. Open source data collection in the developing world. *Computer* **2009**, *42*. [CrossRef]

53. Linchant, J.; Lhoest, S.; Quevauvillers, S.; Semeki, J.; Lejeune, P.; Vermeulen, C. WIMUAS: Developing a tool to review wildlife data from various UAS flight plans. *Int. Arch. Photogramm. Remote Sens. Spat. Inf. Sci. ISPRS Arch.* **2015**, *40*, 379–384. [CrossRef]

54. Macmillan, N.A.; Creelman, C.D. *Detection Theory: A User's Guide*; Laurence Erlbaum Associates Inc.: Mahwah, NJ, USA, 2005.

55. Bates, D.; Maechler, M.; Bolker, B.; Walker, S. lme4: Linear mixed-effects models using Eigen and S4. *R Package Version* **2014**, *1*, 1–23.

56. Crawley, M.J. *The R Book*; John Wiley & Sons, Ltd.: Hoboken, NJ, USA, 2017.

57. López-Bao, J.V.; Rodríguez, A.; Palomares, F. Behavioural response of a trophic specialist, the Iberian lynx, to supplementary food: Patterns of food use and implications for conservation. *Biol. Conserv.* **2008**, *141*, 1857–1867. [CrossRef]

58. Ancrenaz, M.; Gimenez, O.; Ambu, L.; Ancrenaz, K.; Andau, P.; Goossens, B.; Payne, J.; Sawang, A.; Tuuga, A.; Lackman-Ancrenaz, I. Aerial surveys give new estimates for orangutans in Sabah, Malaysia. *PLoS Biol.* **2005**, *3*, e30003. [CrossRef] [PubMed]

59. van Casteren, A.; Sellers, W.I.; Thorpe, S.K.S.; Coward, S.; Crompton, R.H.; Myatt, J.P.; Ennos, A.R. Nest-building orangutans demonstrate engineering know-how to produce safe, comfortable beds. *Proc. Natl. Acad. Sci. USA* **2012**, *109*, 6873–6877. [CrossRef] [PubMed]

60. Pearse, A.T.; Gerard, P.D.; Dinsmore, S.J.; Kaminski, R.M.; Reinecke, K.J. Estimation and correction of visibility bias in aerial surveys of wintering ducks. *J. Wildl. Manag.* **2008**, *72*, 808–813. [CrossRef]

61. Chabot, D.; Bird, D.M. Evaluation of an off-the-shelf unmanned aircraft system for surveying flocks of geese. *Waterbirds* **2012**, *35*, 170–174. [CrossRef]

62. Stewart, F.A. *The Evolution of Shelter: Ecology and Ethology of Chimpanzee Nest Building*; University of Cambridge: Cambridge, UK, 2011.

63. Hicks, T.C. A Chimpanzee Mega-Culture? Exploring Behavioral Continuity in Pan Troglodytes Schweinfurthii Across Northern DR Congo. Ph.D. Dissertation, Universiteit Van Amsterdam, Amsterdam, The Netherland, 2010.

64. Van Schaik, C.P.; Wich, S.A.; Utami, S.S.; Odom, K. A simple alternative to line transects of nests for estimating orangutan densities. *Primates* **2005**, *46*, 249–254. [CrossRef] [PubMed]

65. Gonzalez, L.F.; Montes, G.A.; Puig, E.; Johnson, S.; Mengersen, K.; Gaston, K.J. Unmanned aerial vehicles (UAVs) and artificial intelligence revolutionizing wildlife monitoring and conservation. *Sensors* **2016**, *16*, 97. [CrossRef] [PubMed]

66. Selby, W.; Corke, P.; Rus, D. Autonomous aerial navigation and tracking of marine animals. In Proceedings of the Australasian Conference on Robotics and Automation, Melbourne, Australia, 7–9 December 2011; pp. 1–7.

67. Abd-Elrahman, A.; Pearlstine, L.; Percival, F. Development of pattern recognition algorithm for automatic bird. *Surv. Land Inf. Sci.* **2005**, *65*, 37.

68. Hodgson, J.C.; Mott, R.; Baylis, S.M.; Pham, T.T.; Wotherspoon, S.; Kilpatrick, A.D.; Segaran, R.R.; Reid, I.; Terauds, A.; Koh, L.P. Drones count wildlife more accurately and precisely than humans. *Methods Ecol. Evol.* **2018**, *1*, 1–19. [CrossRef]

69. Andrew, M.E.; Shephard, J.M. Semi-automated detection of eagle nests: An application of very high-resolution image data and advanced image analyses to wildlife surveys. *Remote Sens. Ecol. Conserv.* **2017**, *3*, 66–80. [CrossRef]

70. Duffy, J.P.; Anderson, K. A 21st-century renaissance of kites as platforms for proximal sensing. *Prog. Phys. Geogr. Earth Environ.* **2016**, *40*, 352–361. [CrossRef]

71. Du, T.; Schulz, A.; Csail, M.; Zhu, B.; Bickel, B.; Matusik, W. Computational multicopter design. *ACM Trans. Graph.* **2016**, *35*. [CrossRef]

72. Magnussen, Ø.; Hovland, G.; Ottestad, M. Multicopter UAV design optimization. In Proceedings of the 2014 IEEE/ASME 10th International Conference on Mechatronic and Embedded Systems and Applications (MESA), Senigallia, Italy, 10–12 September 2014. [CrossRef]

73. Berni, J.A.J.; Member, S.; Zarco-tejada, P.J.; Suárez, L.; Fereres, E. Thermal and narrowband multispectral remote sensing for vegetation monitoring from an unmanned aerial vehicle. *IEEE Trans. Geosci. Remote Sens.* **2009**, *47*, 722–738. [CrossRef]

74. Gini, R.; Passoni, D.; Pinto, L.; Sona, G. Use of unmanned aerial systems for multispectral survey and tree classification: A test in a park area of northern Italy. *Eur. J. Remote Sens.* **2014**, *47*, 251–269. [CrossRef]

75. Woll, C.; Prakash, A.; Sutton, T. A case-study of in-stream juvenile salmon habitat classification using decision-based fusion of multispectral aerial images. *Appl. Remote Sens.* **2011**, *2*, 37–46.

76. Sugiura, R.; Noguchi, N.; Ishii, K. Remote-sensing technology for vegetation monitoring using an unmanned helicopter. *Biosyst. Eng.* **2005**, *90*, 369–379. [CrossRef]

77. Arnold, T.; De Biasio, M.; Fritz, A.; Leitner, R. UAV-based measurement of vegetation indices for environmental monitoring. In Proceedings of the 2013 7th International Conference on Sensing Technology, ICST, Wellington, New Zealand, 3–5 December 2013; pp. 704–707.

78. De Biasio, M.; Arnold, T.; Leitner, R.; McGunnigle, G.; Meester, R. UAV-based environmental monitoring using multi-spectral imaging. *Proc. SPIE* **2010**, 766811. [CrossRef]

79. Greenwood, F. How to make maps with drones. In *Drones and Aerial Observation: New Technologies for Property Rights, Human Rights, and Global Development*; New America: Washington, DC, USA, 2015; pp. 35–47.

80. Wallace, L.; Lucieer, A.; Watson, C.; Turner, D. Development of a UAV-LiDAR system with application to forest inventory. *Remote Sens.* **2012**, *4*, 1519–1543. [CrossRef]

81. Gooday, O.J.; Key, N.; Goldstien, S.; Zawar-Reza, P. An assessment of thermal-image acquisition with an unmanned aerial vehicle (UAV) for direct counts of coastal marine mammals ashore. *J. Unmanned Veh. Syst.* **2018**. [CrossRef]

82. Hicks, T.C.; Tranquilli, S.; Kuehl, H.; Campbell, G.; Swinkels, J.; Darby, L.; Boesch, C.; Hart, J.; Menken, S.B.J. Absence of evidence is not evidence of absence: Discovery of a large, continuous population of Pan troglodytes schweinfurthii in the Central Uele region of northern DRC. *Biol. Conserv.* **2014**, *171*, 107–113. [CrossRef]

MDPI
St. Alban-Anlage 66
4052 Basel
Switzerland
Tel. +41 61 683 77 34
Fax +41 61 302 89 18
www.mdpi.com

Drones Editorial Office
E-mail: drones@mdpi.com
www.mdpi.com/journal/drones

www.ingramcontent.com/pod-product-compliance
Lightning Source LLC
Chambersburg PA
CBHW051858210326
41597CB00033B/5947